Scientific Irrationalism

Scientific Irrationalism
Origins of a Postmodern Cult

David Stove

Foreword by Keith Windschuttle
Afterword by James Franklin

Transaction Publishers
New Brunswick (U.S.A.) and London (U.K.)

First paperback printing 2007

Copyright © 2001 by Transaction Publishers, New Brunswick, New Jersey. Originally published as *Anything Goes: Origins of the Cult of Scientific Irrationalism* in 1998 by Macleay Press.

This book is printed on acid-free paper that meets the American National Standard for Permanence of Paper for Printed Library Materials.

Library of Congress Catalog Number: 00-048404
ISBN: 0-7658-0063-2 (cloth); 1-4128-0646-1 (paper)
Printed in the United States of America

Library of Congress Cataloging-in-Publication Data

Stove, D.C. (David Charles)
      [Anything goes]
   Scientific irrationalism : origins of a postmodern cult / David Stove ; foreword by Keith Windschuttle ; afterword by James Franklin.
         p. cm.
Originally published: Anything goes. [Australia] : Macleay Publishers, 1998.
Includes bibliographical references and index.
ISBN 0-7658-0063-2 (alk. paper)
   1. Science—Philosophy. 2. Irrationalism (Philosophy). 3. Popper, Karl Ramund, Sir, 1902- 4. Kuhn, Thomas S. I. Title.

Q175 .S827   2000
501—dc21   00-048404

To the memory of
GEORGE ORWELL
who might have enjoyed at least Part One of this book

CONTENTS

FOREWORD

SCIENCE studies has been a feature of academic life in English-speaking countries for the past fifty years. The term 'science studies' refers not to science as it is practised but to inquiries into the nature of science. People in the field have not been scientists *per se* but sociologists of science, historians of science, philosophers of scientific method and, more recently, people who engage in cultural studies of science. In other words, rather than a province of the physical sciences themselves, science studies is a domain within the humanities and social sciences. The field began in the 1940s in the United States, Britain and Australia where departments entitled History and Philosophy of Science were established in some of the leading universities.[1] They had three aims: first, to impart some understanding of science to students taking arts degrees whose education would otherwise be confined entirely to the humanities and social sciences; second, to encourage those learning science to reflect on the intellectual and social implications of their field; and third, to establish centres of research for the study of the development and social impact of science, the nature of scientific inquiry and the philosophical rationale of scientific method. In particular, science studies

1

was initiated to help bridge what came to be famously identified by C. P. Snow in 1959 as the 'Two Cultures' that divided the Western mind: the literary and the scientific.[2] Snow pointed out that not only the average lay person but many who were otherwise highly educated were ignorant about the facts and laws discovered by science, an ignorance complemented by a widespread dearth of understanding of scientific concepts and its methods of research. In the middle of the twentieth century, he noted, most of those educated in the humanities had yet to properly comprehend the scientific revolution of the seventeenth century.

The canonical text of science studies was produced fairly early in the development of the new field. This was *The Structure of Scientific Revolutions* by Thomas Kuhn[3] which was published in 1962 and which has since become one of the most influential books of this century, not only in science studies but in the humanities and social sciences as a whole. In the *Arts and Humanities Citation Index*, a compilation of the references to authors and works made in footnotes to academic papers, *The Structure of Scientific Revolutions* is the most cited single book, on any subject, of this century.[4] Kuhn offered a sociological explanation of how dramatic changes in scientific opinion and methods come about. According to Kuhn, the range of techniques, assumptions and theories used by the members of a particular scientific field can be termed a 'paradigm'. Within a paradigm, researchers practise 'normal science', which is characterised by periods of calm and steady development dominated by one accepted set of concepts. However, normal science is often disrupted by scientific revolutions, such as the overthrow of Ptolemaic astronomy by that of Copernicus, or the replacement of Newton's mechanics by Einstein's theory of relativity. These 'paradigm shifts' occur because anomalies or observations inconsistent with the dominant perspective produce a crisis that eventually leads the scientific community to lose faith in the existing paradigm. The door is then opened for a scientific revolution to occur to establish a new paradigm which explains both the former body of data and the inconsistencies that the old paradigm could not handle. A new period of normal science then continues until it, again, is subject to its own crisis and revolution.

Kuhn also argued for what he called the *incommensurability* of scientific theories. New paradigms may borrow some of the vocabulary and apparatus of the old but they seldom use these borrowed elements in the same way. Different paradigms operate with different concepts, often changing the meaning of old terms, and they have different standards of acceptable evidence, as well as different means of theorising about their subject matter. Einstein's theory of relativity did not add a new increment of knowledge to the secure truth of Newton's theory of gravitation, but overthrew it completely. Moreover, there is no common measure for the merits of competing theories. Hence, there is no way of ranking scientific theories and thus no grounds for arguing that science is progressive. Einstein is not superior to Newton, only different.

One of the reasons for the attention attracted by Kuhn's thesis was his explanation of why scientists came to accept one body of scientific theory against the claims of another. Kuhn insisted that, although a paradigm has to be supported by compelling evidence and arguments in its favour, it is never accepted for purely objective reasons but rather it gains its acceptance because a consensus of opinion within a scientific community agrees to use it. He said the issue is not decided by purely logical argument but is more like sudden conversion or a 'gestalt switch'.[5] The factors that lead scientists to change their allegiance to paradigms, he argued, need to be explained in terms of their values and the personal relations within a scientific community. 'As in political revolutions, so in paradigm choice—there is no standard higher than the assent of the relevant community ... Even the nationality and prior reputation of the innovator and his teachers can sometimes play a significant role.'[6]

Following these developments, a bevy of social scientists entered the field to take up what they saw as one consequence of Kuhn's position: that what is believed in science is determined by custom and power relations. One of the best known of these observers, David Bloor, has argued that the nature of science can be explained through the methods of the sociology of knowledge. Scientists accept scientific laws, he argued, primarily for reasons of justification, legitimation and control.[7] Another frequently cited contribution is the book *Laboratory Life* by Bruno Latour and Steve Woolgar, which purports to provide an anthropological study of the 'internal workings of scientific activity'. Its subtitle, 'The Social Construction of Scientific Facts', plus its conclusion that all of science is merely the 'construction of fictions', led to this position becoming known as constructivism.[8] The sociologist, H. M. Collins, produced a similar study, claiming that scientific experiments never amount to independent evidence for scientific theories. The theories themselves, Collins said, determine what counts as an effective experiment and so there are no objective criteria to separate the outcome of an experiment from the theory it was designed to test. A scientist who accepts another's experimental results, Collins argued, usually does so because of the prestige, nationality and reputation of the experimenter. Hence science is less about discovering the nature of the world than about imposing on the world the institutional relations of the scientific community.[9]

The most radical of those who followed Kuhn was the philosopher Paul Feyerabend who took Kuhn's notion of the incommensurability of scientific theories and used it to argue some notorious conclusions. In his book *Against Method*,[10] Feyerabend applied incommensurability not only to rival theories within science but to the whole of science itself compared to other fields which claim to know the world. Because they, too, are incommensurable, he argued there can be no argument in favour of science over other forms of understanding. He compared science with astrology and voodoo and claimed there is no general criterion that gives scientific knowledge priority over the latter. Hence, he argued, it is wrong to teach science to school children as if it had a monopoly on wisdom. The grip that the ideology of science has on government policy deserves to be broken, he said, in the same way that secular educationalists last century broke the nexus between Church and state. This would clear the way for other approaches, such as magic, to be taught instead of science. 'Thus, while an American can now choose the religion he likes, he is still not permitted to demand that his children learn magic rather than science at school. There is a separation between state and Church, there is no separation between state and science.'[11] In Feyerabend's view, science should be studied not as some holy writ but as a historical phenomenon 'together with other fairy tales such as the myths of "primitive" societies.'[12] Consistent with this line, Feyerabend has defended Christian fundamentalists who want the Biblical version of creation taught in American schools alongside Darwin's theory of evolution. He was not only aware of the logical implications of his case but, borrowing a line from Cole Porter, he jauntily recommended them: 'Anything goes.'[13]

Despite its academic acclaim, Kuhn's thesis attracted a body of critics who pointed out that one of its consequences was that it deprived scientists of the claim to be discovering final truths about the world. Instead, scientific knowledge—that is, empirical, factual knowledge—had to be regarded as relative to the theory or paradigm the scientist inhabited. Kuhn himself dismissed the charge of relativism, but his own writings made his rejection unconvincing. If paradigms are gestalts then scientists on different sides of a Kuhnian revolution will, as he put it himself, live in different worlds. Pre-revolutionary theories will be substantiated by pre-revolutionary evidence and facts, and post-revolutionary theories will be supported by their own post-revolutionary facts.

> Consider ... the men who called Copernicus mad because he proclaimed that the earth moved. They were not either just wrong or quite wrong. Part of what they meant by 'earth' was fixed position. Their earth, at least, could not be moved.[14]

If this were true then Feyerabend's challenge to the rationality of science, including his denial of its special status in determining the truth about the world, would be hard to deny.

However, relativism is a position that most defenders of scientific investigation want to reject. It means that science can no longer be regarded as a universal method for discovering truths. Moreover, it means that any reasonably coherent doctrine or body of beliefs can produce 'truths' of its own. Science is thus reduced to one belief system among many. While the defenders of science may find this distasteful, the 1990s have witnessed the emergence of a group of science critics who actually welcome the news. Academics in the field of cultural studies are now arguing that, rather than a method with universal application, science should be regarded as something that is both culturally and historically limited. It is the product of Western Europe between the seventeenth and twentieth centuries. Other cultures and other times have equally acceptable claims for the product of their intellectual labours. As one of Australia's leading academic sociologists, R. W. Connell, has put it:

The idea that Western rationality must produce universally valid knowledge increasingly appears doubtful. It is, on the face of it, ethnocentric. Certain Muslim philosophers point to the possibility of grounding science in different assumptions about the world, specifically those made by Islam, and thus develop the concept of Islamic science.[15]

The defenders of science usually respond to claims of this kind by arguing that accepting them is no different from supporting some of the more grotesque historical examples of relativism in science: for instance, the conflict between 'Aryan' and 'Jewish physics' that set back German science under the Nazi regime[16], and the claims by the Marxist plant geneticist, T. D. Lysenko, to have developed a 'proletarian' approach to science in opposition to 'bourgeois' science. The application of Lysenko's methods to agriculture not only produced a series of disastrous crop failures in the USSR in the 1930s and 1940s, but was partly responsible for the Chinese famine of 1958–62, the worst in human history, which caused the deaths of between thirty and forty million people during the so-called Great Leap Forward.[17]

While responses like this drawn from anecdote and analogy might appear persuasive to the average lay observer and to most scientists themselves, within the field of science studies they have largely fallen on deaf ears. This is because the case for relativism has proven stubbornly resistant to attack at the level of philosophy and, especially, from the dominant tendency within the philosophy of scientific method. The sociology of science and the philosophy of scientific method are two quite distinct fields. Within the latter, the most influential theorist at the time Kuhn produced his thesis was Karl Popper, whose major work on the subject, *The Logic of Scientific Discovery,* was published in English in 1959.[18] Popper was one of the most honoured British philosophers of this century. He was knighted in 1962, made a Companion of Honour in 1982 and had the rare distinction of election as a Fellow of both the Royal Society and the British Academy. He is best-known for his political writing, especially *The Open Society and its Enemies* and *The Poverty of Historicism*, both critiques of the history of ideas that produced fascism and communism. In the philosophy of science Popper made his reputation as a critic of the school of logical positivism. This was an approach produced by the group known as the Vienna Circle, led by Rudolph Carnap and Carl Hempel, who fled Austria for the USA in the 1930s when their homeland came under Nazi control. By the late 1950s, logical positivism had produced the then most widely-accepted account of science that defended its traditional empirical methodology and its claim to establish universally valid knowledge. Though he also originated in Vienna, Popper had himself long been a critic of the Vienna Circle. The first version of *The Logic of Scientific Discovery* had been published in Austria 1934 as *Logik der Forschung* and was an explicit attack on logical positivism.

The logical positivist school had argued for the view accepted by most scientists that evidence was used to *verify* scientific theories. Traditional scientific method had held since the seventeenth century that we gain scientific knowledge by generalising from our observations, that is, through a process of induction. Popper claimed, however, that the proper role of evidence was to falsify scientific conjectures. Thus, instead of the traditional view that a scientific theory was *verifiable* by observation, Popper contended that a scientific theory is one that is *falsifiable*. Theories, Popper said, were not the kind of things that could be established as being conclusively true in light of observation or experiment. Inductive arguments were invalid. Instead, theories were speculations, guesses or conjectures about some aspect of the world or the cosmos.

The role of observation and experiment was to rigorously test these theoretical conjectures and to eliminate those that failed to stand up to the tests that were applied. Science advanced by trial and error, with observation and experiment progressively eliminating unsound theories so that only the fittest survived. As the title of one of Popper's best known books described it, scientific method is a process of 'conjectures and refutations'[19] in which we learn not by our experience but by our mistakes. Some of Popper's allies took the critique of logical positivism further by questioning its picture of scientific theories emerging from evidence that is composed of raw, empirically given, neutral data. Perception, they argued, is always conditioned by the pre-existing beliefs, desires and culture of the observer, so there can be no such thing as 'raw' data. All observation, they claimed, is theory-laden.[20]

Popper regarded himself as a defender of science and, like most scientists, he was contemptuous of the notion that science was relative to culture. He became a major critic of Kuhn and his thesis for this very reason. The debate between the two dominated discussion in the 1960s and 1970s. Kuhn replied that Popper's own approach was little different to the verification theory it was designed to replace and that Popper's falsificationism did not accord with scientific practice since outright falsifications were rarely found.[21] One of Popper's students, and his successor as Professor of Logic and Scientific Method at the University of London, Imre Lakatos, attempted a kind of reconciliation between the two approaches by introducing the notion of a scientific 'research program', which had similarities to a Kuhnian paradigm. Instead of falsification by observation, Lakatos substituted the contrast between a research program that was progressing and one that was degenerating. A degenerating (rather than falsified) research program was one which no longer made novel predictions compared to a more progressive rival. [22]

Despite Popper's critique of Kuhnian relativism, Popper's own school has itself long been dogged by the same charge. This is because under the falsifiability principle, no scientific theory can ever be conclusive. The most that a scientific theory can ever attain is the status of a conjecture or a hypothesis. We can never have sufficient grounds for gaining from science anything as concrete as 'knowledge' in the usual sense of that word. What this means is that Popper's thesis has great difficulty in explaining why science is superior to any alternative belief system like, say, astrology. A logical positivist could argue that the real difference between astrophysics and astrology is that the findings of the former are established by a large body of evidence while the latter remains merely speculative. However, because he thinks inductive evidence is powerless to substantiate a scientific theory, Popper is in no position to make this kind of distinction. If he is to avoid the charge of relativism, he has to explain why some speculations are better than others, but his theory of falsification is not designed to do this. Unless it has actually been falsified, a system of contingent belief like astrology must remain for Popper a reasonable conjecture. The claim that all observations are 'theory-laden' is in a similar boat. It leads to the conclusion that all observations are necessarily 'subjective', so that observable 'facts' are relative to observers and dependent upon their psychology, their history or their culture.

In recent years, followers of Popper such as Larry Laudan have made a number of attempts both to avoid the charge of relativism and to distance themselves from the Kuhn thesis, especially the extension of it made by Feyerabend.[23] Their efforts, however, have had little impact within what emerged in the 1990s as the major new field within science studies. This is the cultural studies approach to science which, while drawing its rationale from the arguments of Kuhn, Feyerabend and, in some cases, even Popper himself, also has its origins in French poststructuralist and postmodernist literary theory. Cultural studies is not a neutral, disinterested or objective approach to the study of either culture or science. It regards attempts at 'disinterested' or 'objective' research to be naïve since it believes all forms of knowledge to be exercises in power, and hence all scholarship to be political. Cultural studies regards itself as providing the intellectual underpinnings for the various 'identity group' politics that have emerged in the last three decades, especially those of feminism and multiculturalism. From this perspective, the charge of relativism aimed at Kuhn and Popper is not something to be avoided but to be advocated. Since science is something made by people working in groups, it is a social activity put together according to agreed principles. Therefore, the proponents of cultural studies assure us, science is a social construction and its pronouncements are social constructions as well. This thesis opens the way to the claim that different groups of people will do science differently and come to different conclusions. Hence there can be feminist science, or indigenous science or, as noted above, Islamic science, all of which can produce their own 'knowledges' that are just as valid as what were once regarded as the universally applicable findings of Western science. In 1996, the American cultural studies journal, *Social Text*, published a now famous article endorsing this whole approach:

> But deep conceptual shifts within twentieth century science have undermined this Cartesian-Newtonian metaphysics; revisionist studies in the history and philosophy of science have cast further doubt on its credibility; and, most recently, feminist and poststructuralist critiques have demystified the substantive content of mainstream Western scientific practice, revealing the ideology of domination concealed behind the facade of "objectivity". It has thus become increasingly apparent that physical "reality", no

> less than social "reality", is at bottom a social and linguistic construct; that scientific "knowledge", far from being objective, reflects and encodes the dominant ideologies and power relations of the culture that produced it; that the truth claims of science are inherently theory-laden and self-referential; and consequently, that the discourse of the scientific community, for all its undeniable value, cannot assert a privileged epistemological status with respect to counter-hegemonic narratives emanating from dissident or marginalised communities.[24]

This is a passage written by the New York University physicist, Alan Sokal, in an article that has since become known as 'the Sokal Hoax'. The article was a parody of what cultural theorists believe about science, (as well as of the jargon in which they express themselves). It contained a considerable volume of deliberately fallacious claims such as: 'the π of Euclid and the G of Newton, formerly thought to be constant and universal, are now perceived in their ineluctable historicity'. Anyone with a familiarity with high school science should have seen the article was a spoof and the assertions so nonsensical that they were self-evidently untrue. The fact that the editors of *Social Text* failed to recognise it for what it was, and published it in all faith as a serious academic article, demonstrated the paucity of their understanding of the very field of which they had long been critics. Indeed, one of the editors of the journal, Andrew Ross of New York University, had himself published an earlier book called *Strange Weather*, a critique of modern science and technology, which he audaciously dedicated to "all the science teachers I never had. It could only have been written without them".[25]

By the mid–1990s, the cultural studies version of science studies had reached the stage where two of its American opponents, Paul Gross, a biologist, and Norman Levitt, a mathematician, were disturbed enough to put together a lengthy collection of its assertions, under the title *Higher*

Superstition,[26] to demonstrate how far removed from the real world the whole movement had become. Among their targets:

- Feminist theorists like Sandra Harding who, in *The Science Question in Feminism*, calls Newton's *Principia Mathematica* a 'rape manual'. Harding propounds a doctrine she calls 'strong objectivity' which argues for quotas of disadvantaged groups to be applied to research teams and chairs in science on the grounds that, once more women, blacks, gays and lesbians join the ranks, science will become more open and inventive.

- Marxists such as Stanley Aronowitz, whose book *Science as Power* claims that, since science and technology are key elements in the authority and dominance of modern capitalism, the duty of a social critic is to demystify science and to topple it from its position of authority.

- The postmodernist philosopher Steven Best, who calls for the development of 'postmodern science' to counter the 'inherently repressive' nature of modern science. 'Postmodern science,' he argues, would be 'ethically sensitive, spiritually aware, and ecologically sane' compared to traditional science's support for the Western ethos of conquest, domination and objectification.

- The demand for the introduction of 'feminist algebra', justified on the grounds that mathematics at present 'is portrayed as a woman whose nature desires to be the conquered Other.' The authors of this recommendation want the whole field of mathematics reappraised so that masculinist failings can be rectified and it can become a discipline fit for women to enter.

- Textbooks written for the US African-American high school and university systems that claim, on the most fanciful evidence, that ancient Africans, long before the rise of classical Greece, invented aeronautics and had understanding of quantum physics and gravitational theory.

- Academics like Andrew Ross who promote the notion that New Age mystics and others on the far fringes of science are the leading lights in the struggle against the allegedly omnivorous monster of techno-culture.

In other words, from the point of view of traditional science, many of the ideas on the subject that are now commonly expressed within the humanities are truly appalling. Forty years after Snow's diagnosis, the hope that science studies would help bridge the divide between the 'two cultures' has been turned on its head. As the Sokal hoax and the evidence produced by Gross and Levitt demonstrate, much of science studies is now a site not of enlightenment about its subject matter but of political demagoguery, theoretical obfuscation and plain ignorance.

THE MOST effective antidote to all this pretension and confusion is the philosophy of David Stove. In the five compact chapters of the book that follows he demonstrates how extravagant has been the verbiage wasted on this issue and how irrational have been its combatants. Rather than Kuhn and Popper being engaged in debate from the opposite sides of the philosophical divide, Stove shows that they share most of their ground in common. And rather than there being some inscrutable enigma about the justification of the traditional empirical and inductive methods of scientists, Stove argues that the problems have all lain in the reasoning of the critics. As well as the logical mistakes and conceptual elisions made by Kuhn, Popper and their supporters, Stove identifies their collective dependency on a single argument made by the philosopher of the Scottish Enlightenment, David Hume, and then demonstrates how little potency that argument actually has for destabilising the claims of science.

Yet if Stove actually accomplishes all this, why is it that he is not widely recognised as a major player in this debate? Why are the citations of his writings as sparse as those of Kuhn's are plentiful? Partly it is a matter of timing. Stove began his academic career in the 1950s but by the 1960s, when Kuhn's reputation was being made, Stove found himself increasingly at odds with the emerging *zeitgeist*. He was a conservative and anti-Communist at a time when academic life was swept, first, by a wave of protest against the war in Vietnam and a revival of Marxist theory, and then, hard on its heels, by the emergence of identity group politics in the form of feminism and ethnic liberation, which found the university campus an accommodating habitat. In an era of ascending radical hegemony, celebrity status went largely to the critics of the status quo, not its defenders. It is also partly a matter of Stove's loathing of self-promotion. As some of the more caustic comments in this book indicate, he was derisive of philosophers whose reputations derived mostly from their ability to market themselves, who took a popular rather than a rigorous line of argument, and especially of those who put catchy titles on their books. Accordingly, he seemed to have chosen the dullest possible title for the first edition of this work which, when originally published in 1982, was called *Popper and After*. This was almost asking for its readership to be restricted to a small band of *cognoscenti*. Only the subtitle, *Four Modern Irrationalists*, gave a clue to the aggressive argumentation inside.

Another of Stove's handicaps was his inability to hide his contempt for fools and his consequent penchant for making enemies. One of his friends, Peter Coleman, wrote in a 1991 review of Stove's *The Plato Cult*, that although the author was one of Australia's best essayists and polemicists, the book had attracted very few reviewers because it argued that most intellectuals talked nonsense most of the time.[27] So it was not surprising that few scholars and academics were willing to identify themselves with either the author or his work. One reason we can reject, however, is that Stove was handicapped by being located, unnoticed, in the antipodes in the South Pacific. In philosophy, geography today counts for little, as the worldwide reputations won by Australian philosophers as diverse as David Armstrong and Peter Singer demonstrate. Indeed, Stove's *Popper and After* was written in the immediate aftermath of the considerable international acclaim of another book on the philosophy of science written within his own institution, the University of Sydney. This was *What Is This Thing Called Science?* by Alan Chalmers who, when it was published in 1976, was an earnest advocate of the theories of Kuhn and Feyerabend. The success of this book, and especially the wink and nudge to Cole Porter in its title, was one of the sources that most likely irritated Stove enough to write his own.

David Stove died in 1994 and, as so often and so unfairly happens in intellectual history, his reputation has grown considerably since his death. Long before this, though, he had a small circle of admirers, most of whom were academic philosophers, who appreciated not only his intellectual brilliance and the polish of his unadorned prose, but how very funny he invariably was. For instance, Stephen Stich of Rutgers University wrote: 'Stove's essays are elegant, insightful, beautifully crafted and enormously interesting. They are also outrageous, opinionated, occasionally unfair and almost always side-splittingly funny ... He says things that need to be said and that others lack the courage—or foolhardiness—to say'.[28] Michael Levin of the City University of New York said that reading Stove 'is like watching Fred Astaire dance. You don't wish you were Fred Astaire; you are just glad to have been around to see him in action.'[29] Since 1994, the circle of insiders has widened to include many people who had not read Stove when he was alive but who, on discovering him, have asked almost incredulously: why didn't I know of his work before? This new recognition has come partly because of the publication in 1995 of a collection of his (mainly) non-philosophical essays under the title *Cricket Versus Republicanism*,[30] partly because of the standing accorded by his inclusion in 1997 in the *Oxford Book of Australian Essays*,[31] and partly because of a substantial review of his writings in 1997 by Roger Kimball in the New York literary magazine, *The New Criterion*.[32] Moreover, in 1998, Alan Sokal and his physicist colleague, Jean Bricmont, used Stove's arguments as one of the principal supports of their highly publicised critique of French postmodernist theory and relativism in the philosophy of science, *Intellectual Impostures*.[33]

Stove, then, was a man before his time, providing answers to a problem whose eventual, disastrous dimensions were foreseen by very few others when he wrote. Today, however, he has been joined by some equally effective critics, especially real scientists like Gross and Levitt, and Sokal and Bricmont, who have themselves injected some intellectual acumen into a debate that has otherwise constituted a veritable monument to irrationalism. Hence it makes sense to re-publish his work now at a more appropriate time, and with a more appropriate title, for those readers searching for some enlightenment on the issue.

One of Stove's greatest gifts to intellectual life was his ability both to tell when an emperor had no clothes and to point his finger with deadly accuracy at the offending naked body. Science studies deserves to be exposed in this way and *Anything Goes* is a book that reveals this more clearly than most. In another context, the American philosopher, David Lewis, commended Stove as the right man for the task.[34] 'A none-too-gentle shaking does us good,' wrote Lewis. 'Once it was [G.E.] Moore who did the job. Nowadays it is above all Stove, and he does it with devastating wit. Naked emperors beware.'

KEITH WINDSCHUTTLE

AUGUST 1998

PART ONE

PHILOSOPHY AND THE ENGLISH LANGUAGE: HOW IRRATIONALISM ABOUT SCIENCE IS MADE CREDIBLE

CHAPTER ONE

NEUTRALISING
SUCCESS WORDS

MUCH more is known now than was known fifty years ago, and much more was known then than in 1580. So there has been a great accumulation or growth of knowledge in the last four hundred years.

This is an extremely well-known fact, which I will refer to as (A). A philosopher, in particular, who did not know it, would be uncommonly ignorant. So a writer whose position inclined him to deny (A), or even made him at all reluctant to admit it, would almost inevitably seem, to the philosophers who read him, to be maintaining something extremely implausible. Such a writer must make that impression, in fact, unless the way he writes effectively disguises the implausibility of his suggestion that (A) is false.

Popper, Kuhn, Lakatos, and Feyerabend, are all writers whose position inclines them to deny (A), or at least makes them more or less reluctant to admit it. (That the history of science is not "cumulative", is a point they all agree on.) Yet with a partial exception in the case of Feyerabend, none of these writers is at all widely regarded by philosophers as maintaining an extremely implausible position. On the contrary, these are the very writers who are now regarded by most philosophers as giving an account of science more plausible than any other. So if what I have said is true, they must write in a way which effectively disguises the implausibility of their position. My object in Part One of this book is to show how they do it.

Of course I do not suppose that these authors, or even any two of them, agree on every point. Feyerabend argues persuasively, indeed, that in the end Lakatos's philosophy of science differed only in words, not in substance, from his own more openly irrationalist one.[1] And Kuhn had no difficulty in showing the very great amount of agreement that exists between himself and Popper.[2] Lakatos and Popper, on the other hand, are at pains to magnify any distance separating them from Kuhn[3] and would be still less willing to acknowledge affinities with Feyerabend; and Popper is almost equally anxious to distinguish Lakatos's position from his own.[4] To an outside philosopher, indeed, the differences of opinion among the four must appear trifling by comparison with the amount of agreement that unites them. But it is in any case sufficient for my purposes that they all agree so far as to share a certain reluctance to admit the truth of (A).

Everyone would admit that if there has ever been a growth of knowledge it has been in the last four hundred years. So anyone reluctant to admit (A) must, if he is consistent, be reluctant to admit that there has ever been a growth of knowledge at all. But if a philosopher of science takes a position which obliges him, on pain of inconsistency, to be reluctant to admit *this,* then his position can be rightly described as irrationalism or relativism. Lakatos and Popper were therefore right in applying these epithets to Kuhn's position.[5] They were further right, I believe, in the suggestion, which is a major theme running through their comments on Kuhn, that this irrationalism stems from the conflation, in Kuhn's writings about science, of the descriptive with the prescriptive: from his steady refusal to distinguish the history or sociology of science from the logic or philosophy of science.[6]

Kuhn, of course, 'admits the soft impeachment' and defends his practice in this respect.[7] (Feyerabend likewise rejects the distinction between description and prescription.[8]) But Kuhn also retorts that in any case Popper and Lakatos do exactly the same thing themselves.[9] This was a very palpable hit, quite impossible to deny. That he confused the logic with the history of science was a common complaint against Popper, and one only too well-founded, long before Kuhn mentioned it in his *tu quoque;* and to try to defend Lakatos from the same reproach would be even more idle. But if it is true, as these critics of Kuhn alleged, and as indeed it is, that the source of irrationalism in his case is the conflation of the history with the logic of science, then the same cause cannot fail to have the same effect in their own case as well.

The question from which I began may therefore be replaced by a more general one. I asked in effect, "How do these writers manage to be plausible, while being reluctant to admit so well-known a truth as (A)?" But in view of what has just been said we are entitled to ask instead: "How do they manage to be plausible, while being in general so irrationalist as they are? For example, while being reluctant to admit (A)?"

It is easy enough to answer this question, I think, in general terms. The answer lies in what I have just referred to: the constant tendency in all these authors to conflate questions of fact with questions of logical value, or the history with the philosophy of science. That this tendency is present, indeed inveterate, in all these writers is, as I have just indicated, quite widely recognised, and is no more than one could gather, if he could not see it for himself in each of them, from the things they say about one another. And this tendency is a cause sufficient to explain the phenomenon of plausible irrationalism. For it is so powerful in us all, and so productive of confusion where criticism does not check it, that it is easily equal to the task of making irrationalism about science plausible. It has imposed on philosophers grosser absurdities than that before now: for example it enabled Mill to find plausible his 'proof' of the principle of utility. For my own part, at any rate, I have no doubt that this tendency is the main part of the answer, in general terms, to my question.

But it is a deficient answer just because it is in such general terms. What we want explained is a specific phenomenon of literary history: namely that some philosophy of science, which is irrationalist enough to generate reluctance to admit (A), is nevertheless made plausible to thousands of readers who would have no patience at all with an open assertion that no more is known now than in 1580, or that no one ever knows anything. Between such specific facts as this to be explained, and the very general tendency so far offered (and correctly offered, as I think), in explanation of them, there is too wide a gap.

To fill this gap what is required, clearly, is to show in detail how the general tendency to conflate the history with the philosophy of science is carried out in the writings of our authors, in such a way as to disguise their irrationalism and make it plausible. We need a catalogue of the actual literary devices by which this trick is turned. It is this which I attempt to supply.

2

IF you wish to recommend a philosophy of science to readers who are sure to find the irrationalism in it implausible, then your literary strategy must clearly be a *mixed* one. Irrationalism which was open and unrelieved would be found hopelessly implausible. So your irrationalist strokes must be softened, by being mixed with others of an opposite kind, or again by being disguised as themselves of an opposite kind. All our authors, accordingly, employ a strategy which is mixed in this sense;

and in fact many forms of it.

An extreme form of mixed strategy is, simple inconsistency: that is, assert an irrationalist thesis, but also assert others which are inconsistent with it.

Popper furnishes many examples of this, of which the following is one. He staggers us by denying that positive instances confirm a universal generalisation, but reassures us by allowing that negative instances are, as we always thought they were, disconfirmatory (so that for example "(x)(Raven x ⊃ Black x)" is disconfirmed by "Raven a.–Black a", but not confirmed by its negation). For he adopts a criterion of confirmation[10] (one which I have elsewhere called the 'relevance criterion' of confirmation[11]), which is well known to have the consequence that p confirms q if its negation disconfirms q.

A strategy which is mixed in the above sense while falling short of inconsistency, can take the form of stating as the aim of science something which common sense would agree to be at least one of its aims; while also saying other things which imply that it is impossible to achieve this aim.

Popper and Lakatos both do this. They say the aim of science is to discover true laws and theories. But they also say, concerning any law or theory, that because it is universal, its truth is exactly as improbable, even *a priori,* as the truth of a self contradiction[12] : in other words, impossible.

A further form that a mixed strategy can take is this: embrace a methodology which is common–sense as far as it goes, but also say other things which imply that (even if it is possible) it is pointless to comply with it.

Popper does this. He enjoins our utmost efforts to establish empirically the falsity of any proposed law or theory. Yet no labour could be more pointless, if he is right in telling us that (for the reason mentioned in the preceding paragraph) the falsity of any such proposition is already assured *a priori*.

Yet another form that a mixed strategy can take is, of course, equivocation: leave them guessing what it is you really believe, the irrationalist bits, or the other ones.

Kuhn, for example, says that the world is the same after "paradigm-shift" as it was before[13]; that scientists working within different paradigms are nevertheless all studying the same world[14]; etc, etc. Well, of course! He is not some kind of crazy Berkeleian, after all, and these things are just common knowledge, like the proposition (A) from which I began, only more so. But Kuhn also uses every literary means short of plain English to suggest that these things are not so: that on the contrary, the world is somehow plastic to our paradigms.

Of course it is not always easy to tell equivocation from downright inconsistency. Take Kuhn again. Some one may tell me that he is not reluctant, *at all,* to admit my historical truism (A). And certainly it would be easy to point to many passages in his writings which support this interpretation. All those passages, for example, in which he says that normal science, operating under the guidance of a paradigm, *solves problems.* No doubt, in particular, Kuhn would admit that normal science has solved a great many problems since 1580. Well, if it has solved those problems, then those problems have been solved, haven't they? We know Kuhn says that a new paradigm "replaces", "destroys", an old one. But he never says that every solution of a particular problem, achieved under the old paradigm, somehow is "destroyed" or becomes an un-solution under the new. Indeed, how could that be? What would it even mean, to say so? If a problem has been solved then it really has been solved. But if this tautology is not denied, then solutions of problems (unless they were, for example, forgotten) would accumulate through successive paradigms. But what then becomes of Kuhn's famous rejection of the cumulative view of the history of science?

This may be another example, then, of our authors' mixed strategy issuing in an actual inconsistency. But on the other hand it may only be another case of equivocation. When Kuhn speaks of science as having solved problems, he no doubt often uses this phrase in the sense in which people normally understand it: which, whatever it is, may certainly be called an *absolute* sense. But—the idea naturally suggests itself—perhaps he sometimes also uses it in another and weaker sense: one which is more consistent with his repeated assertion that what *constitutes* the solution of a problem is relative to the paradigm, the group, and the time.

This suggestion (although I will not pursue it in connection with the phrase "solving problems") seems to me to furnish the key to the two main literary devices by which our authors make irrationalism about science plausible.

3

THE first of these devices I call *neutralising success-words.* A homely example will explain what I mean.

Nowadays in Australia a journalist will often write such a sentence as, "The Minister today refuted allegations that he had misled Parliament", when all he means is that the Minister denied these allegations. "To refute" is a verb with 'success grammar' (in Ryle's phrase). To say the Minister refuted the allegations is to ascribe to him a certain cognitive achievement: that of showing the allegations to be false. "To deny", on the other hand, has no success-grammar. So a journalist who used "refuted" when all he meant was "denied" has used a success-word, but without intending to convey the idea of success, of cognitive achievement, which is part of the word's meaning. He has *neutralised* a success-word.

When journalists do this, no doubt they mostly do so

inadvertently, from mere ignorance. But imagine the same thing done by a journalist who does know the meaning of the two words, and who believes that in fact the Minister only denied the allegations; but who feels for some reason obliged to use language which, in his own opinion, exaggerates the cognitive achievements of Ministers. (Perhaps the reason is that he thinks his readers will listen to nothing but good about Ministers.) Then we would have what I believe is a very close parallel indeed to the way our authors use language to write about science.

For they use the language of success about science—words importing more or less of cognitive achievement, such as "knowledge", "discovery", "facts", "verified", "understanding", "explanation", "solution (of a problem)", and a great many more besides—they use this language quite as freely as do any of those older historians of science whom they despise. They clearly must do so, at least to some extent, for they would forfeit all plausibility if they were to write about science without ever using any success-words at all. Their substantive philosophy, however, is not really consistent with applying, to science, such words in their ordinary success-implying sense. So while they *use* the language of success, they neutralise it. Not all the time, of course: sometimes they use these words in their ordinary sense, despite the inconsistency involved in doing so. But often enough for such neutralised success words to be a prominent and distinctive feature of the English that they write.

This device is clearly one which, if it *were* used, would help enormously towards making irrationalist philosophy of science plausible. For in this way you can have, as thick as you like on every page, all the optimistic *words* of the old historiography and philosophy of science, reassuring the reader (who needs, after all, to be weaned gradually from whiggish notions of science) while all the time, nothing inconsistent with irrationalism need be being said at all.

I now have to substantiate my suggestion that the device of neutralising success-words is characteristic of our authors.

Before coming to cases it will be worthwhile to notice a passage in which the truth of this suggestion of mine is indirectly admitted at once, by one of our authors himself. This is a remarkable paragraph, occurring early in *Against Method*, in which Feyerabend, who is of course more openly irrationalist than our other authors, tells us that (to put it in my language), whenever he applies success-words to science in that book, they are never to be taken in their ordinary sense, but are *intended* to be always understood as neutralised.

The context was this. Feyerabend has just been expounding his 'anarchist' maxim that anything goes: by which he means that any principle of theory preference (induction, counter-induction, Tarot-card, or whatever) may on a given occasion advance science more than any other would. Then he adds the following:

"Incidentally, it should be pointed out that my frequent use of such words as 'progress', 'advance', 'improvement', etc., does not mean that I claim to possess special knowledge about what is good and what is bad in the sciences and that I want to impose this knowledge upon my readers. *Everyone can read the terms in his own way* and in accordance with the tradition to which he belongs. Thus for an empiricist, 'progress' will mean transition to a theory that provides direct empirical tests for most of its basic assumptions. Some people believe the quantum theory to be a theory of this kind. For others, 'progress' may mean unification and harmony, perhaps even at the expense of empirical adequacy. This is how Einstein viewed the general theory of relativity. *And my thesis is that anarchism helps to achieve progress in any one of the senses one cares to choose.* Even a law-and-order science will succeed only if anarchistic moves are occasionally allowed to take place."[15]

It is surely obvious that this addendum to the anarchist methodology is not (what it seems meant to be) an addition which makes that methodology still more permissive. It just makes it totally pointless. We should perhaps think well of a man's heart, if he gives a million-dollar prize for an advance towards a cure for cancer, and says that anything goes as to what means (scientific, magical, or any other) are taken to that end. But if he adds that, 'incidentally', anything goes, too, as to what *counts* as an advance towards a cure for cancer—that "everyone can read these terms in his own way"—then it will be impossible to think well of his head. It is not as though the second piece of permissiveness is an extension of the first: it simply takes all point out of it.

For my purposes, however, the main importance of the passage is this. It is an admission that Feyerabend's philosophy of science, if it were to be consistently expressed, would require that the success-implication of words like "knowledge" and "discovery", as well as of the weaker success-words he mentions himself, be always taken out. It is therefore a strong advance indication that at least in his philosophy we will find that those implications are often taken out; that is, that such words will be neutralised. And since what requires their neutralisation there is the irrationalism he shares with our other authors, it is also an indication that they too will be able to be caught neutralising success-words.

To be sure, Feyerabend does not do what he said he would. Having undertaken to neutralise all success-words, he promptly forgets all about his undertaking, and when it suits him, as it often does, writes about the history of science like any mere Sarton, Wolf, or Pledge. "It is now known that the Brownian particle is..."[16] etc., etc. That is, he often uses words like "known" with their ordinary success-grammar. This was to be expected.

It is just another instance of that mixed strategy which all our authors are obliged, as I have said, to employ.

But to come to details.

One way to neutralise a success-word is to put it in quotation-marks. Thus, in certain circumstances a journalist might write "The Minister 'refuted' the allegations", meaning, and being understood to mean, that the Minister did not refute but only denied them. This might be thought a device too unsubtle for authors such as ours to have made use of. It is not so, however. In any case some variations on the device are not altogether without subtlety. One such variation is what I call "suspending" success-grammar: putting a success-word in quotation marks, not necessarily in order to neutralise it, but just with the intention, or at least the effect, of leaving the reader *uncertain* whether you have neutralised it or not. (This is the effect momentarily produced by signs advertising 'fresh' fish.) Another variation is, using the same success-word several times in close succession, and sometimes putting it in quotation-marks and sometimes not, but with no reason that the reader can discover for so doing. Such variations as these can achieve, partially or gradually, that separation of a success-word from its success-meaning, which quotation-marks sometimes achieve completely and abruptly. They are devices, therefore, which are not at all too unsubtle, nor yet too subtle, to be of some use to a philosopher interested in making irrationalism about science plausible. It would be no use for such a philosopher, and everyone now knows it would be no use, to cry "stinking fish" about science. But it may well be some use for him to *praise* science as "'fresh' fish"; especially if he does it often enough.

Lakatos has certainly done it often enough. Enclosing success-words in quotation-marks was in fact a kind of literary tic with him. He could scarcely have gone to more extravagant lengths in the use of this device, if he had been trying to bring it into disrepute; which, however, he certainly was not.

Take his *Proofs and Refutations*. The first word in this title is of course a success-word. In the book it is subjected countless times to neutralisation or suspension of its success-grammar by quotation-marks. Often, of course, perhaps equally often, Lakatos uses the word without quotation-marks. But what rule he goes by, if he goes by any rule, in deciding when to put quotation-marks around "proof" and when to leave them off, it is quite impossible for a reader of that book to discover. Nor does the reader know what meaning the writer intends to leave in this success-word. He knows that the implication of success is often taken out of it; or rather, he knows that on any given occurrence of the word in quotation-marks, this implication may have been taken out of it. But what meaning has on those occasions been left in it, he is entirely in the dark. Indeed, by the end of the book, or even half-way through it, the reader no longer dares attach success-grammar to "proof" or "proved", even when they occur without quotation-marks. Will any reader of *Proofs and Refutations* undertake to say what the first word of the title means in that book?

By the time Lakatos came to write about empirical science, his tic had got worse. I draw an example from 'Falsification and the Methodology of Scientific Research Programmes'. One short example will suffice, because Lakatos's English is everywhere much the same, and anyone familiar with it will recognise in the following a representative specimen of it.

"One typical sign of the degeneration of a programme which is not discussed in this paper is the proliferation of contradictory 'facts'. Using a false theory as an interpretative theory, one may get—without committing any 'experimental mistake'—contradictory factual propositions, inconsistent experimental results. Michelson, who stuck to the ether to the bitter end, was primarily frustrated by the

inconsistency of the 'facts' he arrived at by his ultra-precise measurements. His 1887 experiment 'showed' that there was no ether wind on the earth's surface. But aberration 'showed' that there was. Moreover, his own 1925 experiment (either never mentioned or, as in Jaffe's [1960] misrepresented) also 'proved' that there was one (cf.. Michelson and Gale [1925] and, for a sharp criticism, Runge [1925])."[17]

Here, in the space of seven lines of print, Lakatos manages to neutralise by quotation-marks three success-words, two of them twice each: "facts", "showed", and "proved".

The effect on the reader is characteristic. An episode in the history of science has been described to him, and it is described, as we see, entirely in words importing cognitive achievement. Yet by mere dint of quotation-marks, every single implication of cognitive achievement has at the same time been neutralised or suspended. The reader, remember, almost certainly has no such knowledge of his own of the episode as would enable him to object, for example, that Michelson really did show one of the things that Lakatos says he "showed". Nor has the reader any idea, as I said before, how much if anything of the ordinary meaning of the various success words the writer is leaving in them: he only knows that their success-implication has been, or may have been, taken out. What, then, will the reader be able to carry away from this passage? Nothing at all; except a strong impression that despite all the success-words used in describing it, there was, in this presumably representative episode from the history of science, no cognitive achievement whatever.

This passage is a very model of irrationalist philosophy of science teaching by example, and being made plausible by example. Yet it depends entirely for its effectiveness on a device at first sight so trivial as the use of quotation-marks to neutralise success-words.

Where Lakatos raises storms of neutralising quotation-marks, Feyerabend, in *Against Method,* just keeps up a steady drizzle of them. For this reason short passages cannot be quoted from him to such effect as they can be from Lakatos. Feyerabend, as we saw, does not keep his promise to neutralise all success-words, but still he often does neutralise them; and when he does it is often by means of quotation-marks. The word "facts", for example, is often thus neutralised: for example, on pp. 40, 41, 46, 47. But he does not neutralise only strong success words. Any success-word, however weak its success-implication may be, or any word which has even an indirect connection with cognitive achievement, he is likely to sprinkle with this soothing balm. For example, "success", p. 44; "truth", pp. 28, 171; "progress", pp. 27, 296; "objective", pp. 19, 181; "rational", pp. 154, 190, 198. There will be no hint left in science of anything as hurtful and undemocratic as success, if Feyerabend can help it.

Popper has always made a certain amount of use of quotation-marks for neutralising success-words. It is well-known, for example, that though he has always been sure that scientific theories can be disconfirmed, he is still not sure, after fifty years, whether our best-confirmed theories are confirmed, or only 'confirmed'.[18] He knows that when he puts quotation-marks around "confirmed", he incurs the suspicion of Humean irrationalism. But then, if he leaves them off it, people may suspect him of believing in non-deductive logic. Hence his indecision.

Kuhn hardly ever resorts to quotation-marks when he wants to neutralise a success-word.

The easiest way, however, to neutralise a success-word, is—just to do it: "bald neutralising", I will call it. That is, just to use a word which implies cognitive achievement, as though it did not. Set at defiance all mere logicians, Oxford philosophers, accurate speakers, and pedants generally. It is even money, after all, whether your solecism will even be noticed; and with luck it may even catch on.

For the sake of plausibility, of course, you should not do this all the time. At any rate not all the time to all those words which, like "knowledge", have the strongest implication of success and are at the same time nearly indispensable for writing about science. If a word is comparatively dispensible, or has comparatively weak success-grammar, you may be able to get away with baldly neutralising it every time.

Bald neutralising is, in Lakatos, subordinated to his main weapon for the destruction of scientific success, the quotation-mark. But in all our authors it is common, and is one of the distinguishing features of their English. I will shortly prove this by examples, in connection with two of the strongest success-words, namely "knowledge" and "discovery". And if this can be done then I can fairly be excused, I think, from documenting the execution our authors do on the weaker and more defenceless members of the success-tribe: "confirmation", "explanation", "understanding", "scientific progress", and the like. The execution is terrific, as may be imagined. When the most emphatic of success-words, such as "knowledge", can be murdered with impunity in open day, as they are by our authors, then the quiet extinction of weaker ones will never attract criticism, or even attention. But to document this process in detail would clearly take far too long.

First, then, Kuhn on knowledge. He says that on the cumulative view, "in the evolution of science new knowledge would replace ignorance"[19], but that this is quite wrong. What really happens is that one paradigm replaces another, and then "new knowledge ... replace [s] knowledge of another and incompatible sort."[20] Kuhn writes, therefore, as though some knowledge can be incompatible with other knowledge; and indeed, on his views, such incompatibility must not only be possible, but common in the history of science and even of the essence of it. It is not possible, however: this is just baldly neutralising the word "knowledge". Knowledge implies truth, and truths cannot be incompatible with one another.

Again, Kuhn simply takes the truth-implication out of the word "knowledge" when he writes, for example, in his most overtly relativist vein, that every scientific theory now discarded (such as Ptolemaic astronomy) possessed in its heyday "the full integrity of what we now call sound scientific knowledge."[21]

The word "discovery", too, Kuhn baldly neutralises at his pleasure. To discover what is not true, or what does not exist, is certainly no mean feat; or rather, it is a simple logical impossibility, forbidden by the success-grammar of the verb "to discover". Yet the history of science as Kuhn recounts it, contains "discoveries" of what is not true; and again, such things must be in fact extremely common on his views. Here is one example, not the only one. "Given Galileo's paradigms, pendulum-like regularities were very nearly accessible to inspection. How else are we to account for Galileo's discovery that the bob's period is entirely independent of amplitude, a discovery that the normal science stemming from Galileo had to eradicate and that we are quite unable to document today."[22]

Feyerabend's promise to neutralise all his success-words in *Against Method* is carried out baldly enough in some cases. For example, on the word "facts" on pp. 29-33, and on the word "knowledge" in the following representative passage.

"Knowledge so conceived is not a series of self-consistent theories that converges towards an ideal view; it is not a gradual approach to the truth. It is rather an ever increasing *ocean of mutually incompatible (and perhaps even incommensurable) alternatives,* each single theory, each fairy tale, each myth that is part of the collection forcing the others into greater articulation and all of them contributing, via this process of competition, to the development of our consciousness."[23]

The "so-conceived" in the first line here means, "as I, Feyerabend, conceive it". Words meaning what they do, however, his 'conception' is mere nonsense. It may be true, or at least intelligible, to say that 'knowledge' is an ocean of incompatible etceteras, or that what passes for knowledge is an ocean of incompatible etc. But it makes no sense to say that knowledge *is* an ocean of incompatible etc., or even (what presumably Feyerabend meant) that the objects of knowledge are an ocean of incompatible etc. Knowledge entails truth, and truth entails possible truth, and possible truth entails compatibility. These are facts about the meaning of common English words, and facts which are, in themselves, not especially important. They *are* facts, though, and because they are, you might as well say that knowledge is a poached egg, as say what Feyerabend says about it here.

All our authors except Popper, it should be understood, not only exercise but more or less openly claim the right to talk nonsense. Feyerabend would exclude no one from this right. (But then he is all heart and "would not hurt a fly".[24]) He thinks that talking nonsense is just good for you, like many other things which are familiar to us all, and the value of which no well-disposed person denies: like rotation of crops, state control of scientists, and turning yourself into a wolf and back again. Lakatos is far more exclusive. Talking nonsense, when it is done by people he approves of, he calls "language-breaking", and he hints that all the very best people do it. Certain great scientists, he implies, have possessed this gift for language-breaking,[25] and it would be a dull reader indeed who could not name one other person that Lakatos thinks is gifted in the same way. Kuhn in his more demure style merely warns us in his Introduction that what he says "strains customary usage"[26] which, when you think about it, is at any rate not more than the truth.

Yet I have no doubt that Kuhn and Lakatos (Feyerabend may be different) would react just as any other philosopher would, if they were told by some one else, such as a mere undergraduate in an essay, that knowledge is a poached egg; or, say, that knowledge entails falsity; or that belief entails knowledge. But what could they or anyone say to such a student, except that, words meaning what they do, what he has written either makes no sense at all, or at the best is necessarily false? And that is what we must often say about what *they* write.

For sheer bald neutralising of success-words, however, Popper remains in a class of his own. It is reasonable to believe, indeed, in view of his extensive influence on our other three authors, that it was from him that they learned what skill they have in this art. Anyway Popper has left monuments of the art which are not likely ever to be excelled. He actually seems to prefer neutralising the very strongest of success words, and to prefer to do it as publicly as possible: that is, in the very titles of his books and articles.

The title of his most famous book in its English translation is a uniquely daring instance of the use of the old optimistic language of the historiography and philosophy of science (the rationalistic and authoritarian language, Lakatos and Feyerabend would say) to introduce a book which, by its actual contents, did far more than any other intellectual cause to discredit that language, and to inaugurate the irrationalist revolution in the historiography and philosophy of science. Thousands of readers have noticed this fact, so far as it concerns the use in that title of the word "logic"; and even Lakatos remarks upon the "paradoxicality"[27] of the title in that respect. But my present concern is with the other part, because of the success-word it contains.

"The Logic of Scientific Discovery", indeed! There is scarcely a word in it, or in anything else Popper ever wrote, about scientific discovery, and the reason is as simple as it is sufficient. "Discovery" is a *success-word,* and of the strongest kind: it means the same as "discovery of what is true or of what exists". The history of science, therefore, to the extent that it *has* been a history of discovery—as it has been so markedly in the last four hundred years, for example—is a history of *success.* But that is not the way that Popper sees the history of science, far from it. For him the history of science is a succession of 'problems', 'conjectures and refutations', Socratic or Pre-Socratic dialogues, 'critical discussions'. It is all talk. In this context any vivid reminder of an actual scientific discovery would be as out of place as a hippopotamus in a philosophy class. The only thing worse would be a reminder (though this would be *too* horrible) of what whig historiography used so often to bracket with scientific discoveries: inventions. Popper is perhaps the first person to see, in the glorious history of scientific discovery, nothing more productive and exhilarating than a huge W.E.A. philosophy class, and one which, to add to its charms, might go on forever. Does anyone suppose that Popper ever wrote or meant to write a book for which a non-misleading title would have been "The Logic of Scientific Discovery of Truth, or of what Exists"? Yet that is a purely analytic extension, only objectionable on aesthetic grounds, of his actual title. But clearly *this* title would belong, in the history of thought about science, in the heyday of the 'whig supremacy', probably somewhere between J. S. Mill and Samuel Smiles, and it sounds a good deal more like the latter than the former.

No, the right title for that book—and it is of some importance to realise that I am here only saying what everyone familiar with its contents has been at least half conscious of all along—would have been "The 'Logic' of Scientific 'Discovery'". But of course *that* would have been too openly irrationalist. Better to let the word "discovery" stand, and trust to the contents of the book, rather than to quotation marks in the title, to neutralise the unintended implication of success. Which duly happened, and never a word said.

It is the word "knowledge", however, which was the target of Popper's most remarkable feat of neutralisation. This word bulks large in his philosophy of science (much larger than "discovery"), and in recent years, in particular, the phrase "the growth of knowledge" has been a favourite with him and with those he has influenced most. Some people have professed to find a difficulty, indeed, in understanding how there can be *growth*-of-knowledge and yet no *accumulation*-of-knowledge. But then some people cannot or will not understand the simplest thing, and we cannot afford to pause over them. Let us just ask, how does Popper use the word "knowledge"?

Well, often enough, of course, like everyone else including our other authors, he uses it with its normal success-grammar. But when he wishes to give expression to his own philosophy of science he baldly neutralises it. Scientific knowledge, he then tells us, is "conjectural knowledge". Nor is this shocking phrase a mere slip of the pen, which is what anywhere else it would be thought to be. On the contrary, no phrase is more central to Popper's philosophy of science, or more insisted upon by him. The phrase even furnishes, he believes, and as the title of one of his articles[28] claims, nothing less than the "solution to the problem of induction".

In one way of course this is true, and must be true, because any problem clearly must yield before some one who is prepared to treat language in the way Popper does. What problem could there be so hard as not to dissolve in a sufficiently strong solution of nonsense? And nonsense is what the phrase "conjectural knowledge" is: just like, say, the phrase "a drawn game which was won". To say that something is known, or is an object of knowledge, implies that it is true, and known to be true. (Of course only 'knowledge that' is in question here.) To say of something that it is conjectural, on the other hand, implies that it is not known to be true. And this is all that needs to be said on the celebrated subject of "conjectural knowledge"; and is a great deal more than should need to be said.

In all our authors there is another misuse of language, and one which is even of an opposite kind to that of neutralising success-words, the explanation of which is nevertheless furnished by that very process. The most striking instance of it is Popper's misuse of the word "guess".

He says that "we must regard all laws and theories as guesses."[29] Taken on its own this would be an inexplicable thing for anyone to say. For who is so ignorant, or so irrationalist, as to believe that? Recall what a guess is. A paradigm case of guessing is, when captains toss a coin to start a cricket match, and one of them 'calls', say "heads". This cannot be a case of knowledge, scientific knowledge or any other, if it is a case of guessing. If the captain knows that the coin will fall heads, it is just logically impossible for him also to *guess* that it will. More than that, however: guessing, at least in such a paradigm case, does not even belong on what may be called the epistemic scale. That is, if the captain, when he calls "heads", is guessing, he is not, in virtue of that, believing, or inclining to think, or conjecturing, or anything of that sort, that the coin will fall heads. And in fact, of course, he normally is not doing any of these things when he guesses. He just calls. And this is guessing, whatever else is.

Now, does Popper believe that what he calls "the soaring edifice of science"[30] is built out of cricket captains' calls of "heads" and "tails"; or of other things in the same epistemic, or rather non-epistemic, boat? Presumably not. Then what has happened?

Simply this. If, when you talk about science, you insist on neutralising success words, depressing them (as it were) on the epistemic scale, then in the interests of plausibility you will find yourself obliged, as if by a kind of hydraulic compensation, to *elevate* in the epistemic scale some non-success-words, and even to promote onto the epistemic scale other words which do not belong on it at all. The former is what has happened to some extent to the word "theory" in all our authors. The latter is what has happened to the word "guess" in Popper.

This compensatory process is not confined to philosophers, but appears to have already affected for the worse the language of scientists themselves. An especially common instance is this: a scientist will say that p *is consistent with* q, when what he means, and is understood by other scientists to mean, is that p *confirms* q. For example, when what he means and is understood to mean is that the red-shift of light from other galaxies confirms the hypothesis that those galaxies are receding from ours, he will say instead that the red-shift is consistent with galactic recession. The absurdity of such a remark, if "consistent with" has its usual sense, is only too evident. Obviously red-shift is consistent with the hypothesis of galactic recession; it is consistent with the absence of galactic recession too; almost every proposition, come to that, for example "Socrates is mortal", is consistent with galactic recession; and with its negation. But it is evident enough, too, what the pressure is, that leads to such statements being made. The influence of Popper's irrationalist philosophy in particular, and of 'modern nervousness' in general, is so widespread and powerful, that even scientists now prefer to steer clear of a word which is even so weakly suggestive of cognitive success as "confirms". So, although "confirms" is what they mean and are understood to mean, they say instead "is consistent with".

This phenomenon, although it is of course yet another abuse of language by our authors and those whom they influence, is one which, as far as it has gone, is rather encouraging than otherwise. For it suggests that, at least until the final triumph of irrationalist philosophy of science, language will to some extent obey a 'law of the conservation of success-grammar'. That is, if you empty all success grammar out of certain words, some of it is going to seep into other words, including some which were quite devoid of it before. Still, this process has not gone very far up to the present, and is not likely to go further than it already has. On the whole our authors' efforts to eradicate belief in scientific success have been remarkably successful.

I N all our authors except Popper there is yet another process which is a natural complement to that of neutralising success-words, or rather is a simple logical extension of it. This is the neutralising, though in the opposite direction as it were, of *failure-words* when they are applied to science: words like "error", "mistake", "is refuted", "is falsified", and so on. With failure-words, or at least with the strongest ones such as those just mentioned, neutralisation typically consists, of course, in removing the implication, which is part of the meaning of such words, of the falsity of the proposition in question.

This kind of neutralisation can be done, just like the other, by means of quotation-marks, or again baldly. It is done in both ways by our authors. For an example of the former the reader need only look back to the quotation to which footnote

17 above is appended. There he will see, what by now he could predict, that when Lakatos used the phrase "experimental mistake", he was quite unable to keep his quotation-marks off it.

With failure-words as with success-words, it was Popper who showed the way ahead. For he laboured long to persuade scientists that no professional stigma attaches to their being refuted. Nor did he labour in vain, but rather to such effect that he succeeded in persuading some of the sadder Popperian scientists that to be refuted was actually the goal of all their endeavours. (They appear to have had rather successful careers.) Yet Popper had only a Pisgah-view of this matter, because he never neutralised the implication of falsity in saying, not of a scientist but of a *proposition,* that it "is refuted" or "has been falsified". On the contrary, that implication was for him the whole point of such phrases.

To take this great leap forward has been left to Lakatos and Feyerabend. As other public benefactors removed the social stigma from illegitimacy, these two reformers have removed the stigma of falsity from refuted propositions. It is true that in these early days (just as happened when Popper first neutralised success-words in the very first days of the revolution), irreconcilable bourgeois elements complain that they do not know what meaning is left in "refuted" and the other strong failure words, now that the implication of falsity has been taken out of them. Well, no doubt these temporary shortages of meaning are unfortunate; but then, a revolution in the philosophy of science is not a tea-party. That the falsity-implication *has* been taken out of failure-words, however, the reader can satisfy himself on scores of pages of Lakatos and Feyerabend. Or if the reader prefers to take his instruction in newspeak neat, here it is in so many words: "If a theory is refuted, it is not necessarily false.[11] For these authors at least, the era of "falsism", as the odious old practice of discrimination against refuted propositions is now called (or soon will be), is over.

On failure-words Kuhn is, as he usually is, less open, and far more chilling, than even Lakatos and Feyerabend.

If the reader looks back to the quotation to which footnote 22 above is appended, there is one thing he can hardly fail to notice: the distance that Kuhn will go about in order to avoid saying that some one working with a paradigm since replaced, as Galileo was, believed, because of that paradigm, something which is *not true*. Kuhn himself draws attention, three pages from the end of his book,[32] to the fact that up to that point he had not once used the word "truth". This avoidance seems understandable. "Truth" is the most stripped-down of all success-words, and therefore does not lend itself to being neutralised in its turn. So the only thing an irrationalist philosopher of science can do with it is to avoid it as long as he can.

But wait a moment: *is* this so understandable? Kuhn is as lavish as the next man in applying to science words like "knowledge" and "discovery", which all import truth anyway: so what can be the point of this fantastic punctiliousness about avoiding the *word* "truth"? But you need only to recall, first, that along with our other authors Kuhn is engaged in taking the success-implication out of those success-words which he does not simply avoid; and second that, on pain of losing all plausibility, you cannot take the success-grammar out of, or avoid, all the success words all of the time. Then this puzzling fragment of literary history will become entirely understandable once more.

But Kuhn is not content to avoid expressions like "not true" and "false" himself. When one of our other authors permits himself to employ, in connection with scientific theories, words which imply falsity, Kuhn is up in arms against him. And for what? Why, for his *misusing language!* For his "odd usage" of words; for saying something "difficult to understand"; for perpetrating a "gross anachronism"; for permitting questions to be raised which are not even "sensible". Truly, there is nothing in any of our authors, not even Feyerabend's belief in lycanthropy, more staggering than this.

The quotations in the preceding paragraph, and in the present one, are all from Kuhn [1970a] pp. 10-13, where he is criticising Popper for using, in connection with science, phrases such as "trial and error" and "learning from our mistakes". "Mistake" and "error" both, you see, import falsity. Kuhn thinks that to call an out-of-date scientific theory, such as the Ptolemaic astronomy, "mistaken" or "a mistake", will "immediately seem an odd usage." Kuhn says: "it is difficult for me to understand what Sir Karl has in mind when he calls that system, or any out-of-date theory, a mistake;" Indeed, Kuhn says, to call *any* out-of-date theory mistaken would be "so gross an anachronism" that no one of "sound historic instinct" is likely to be guilty of such a lapse. To speak in that way, he says, is to invite questions which are not even "sensible", such as: "What mistake was made, what rule broken, when and by whom, in arriving at, say, the Ptolemaic system?"

Was there ever effrontery, if that is what this is, at once so bold and so hollow? Or a challenge so easily met?

Here is one of the many mistakes which was made in arriving at the Ptolemaic system: the belief that the sun goes round the earth every day. As to propriety of language, consider the sentence: "The Ptolemaic system of astronomy is false". There is no 'odd usage' in that: it is good common English. Nor is there any anachronism in it. (One cannot help wondering what Kuhn thinks the word "anachronism" means.)

What this sentence says is, quite obviously, what Popper 'had in mind' when he called Ptolemaic astronomy a mistake. What it says is, moreover, true and well-known to be so. And if Kuhn really believes that no out-of-date scientific theory—no theory, that is which was accepted once but is not now—can properly be called mistaken, then he must also accept the consequence: that every scientific theory which *can* properly be called mistaken is accepted now if it ever was. A consequence none too consistent with that 'growth of knowledge' which he, like our other authors, is always willing to acknowledge, at least in words.

We should for a moment try, though it is almost impossible, to take in the full grotesqueness of the contemporary situation in the philosophy of science. We have already encountered Popper, a grown man and a professor, implying that it is a guess—that is, *something like a cricket captain's call of "heads"*—that the sun does not go round the earth every day. But here is Kuhn, perhaps the most learned and certainly the most influential of living historians of science, writing in such a way as to imply that, like a great many people in 1580 and a few uncommonly ignorant ones even now, he *does not know that it is false* that the sun goes round the earth every day! And implying too, what is far worse still, that to say that he or anyone else *does* know this would be a glaring misuse of language!

I am sure that this was not effrontery. It is simply a revelation, and all the more terrifying for having been made inadvertently, that Kuhn has simply lost contact with the meaning of common English words (such as "false"), and now knows *only* the vocabulary of his own irrationalist philosophy of science.

This must be very close, at least, to the end of the line. Non-cognitivist philosophies of morals are one thing; but here we have a non-cognitivist philosophy of *science*. Our philosophy of science, as I remarked earlier, lost contact long ago, at least as early as Popper, with the refreshing realities of scientific discovery and invention: with the actual objects of science. But with Kuhn even the *intensional* objects of science, the propositions of science, have vanished into thin air, and with their disappearance, of course, the *cognitive* aspect of science vanishes too. Science, it turns out, whatever may be believed to the contrary by the vulgar and by whig historians, is really as intransitive as sleep.

This conclusion is willingly embraced or at least is implied, by all our other three authors. Thus Feyerabend loves to write such phrases as "science, religion, prostitution and so on" [33] and says that science has no more "authority [cognitive or other] than any other form of life."[34] The same non-cognitive conception of science is also what allows Feyerabend to demand that scientific laws be put to the vote;[35] and again is what authorises him, in the middle of mystifying his readers about science, or rather, as a part of that process, to bore them about art.[36] The same conception of science as having no cognitive aspect, nothing to do with knowledge or belief, is implied by Popper and Lakatos. For while knowledge entails belief, they both insist that science has nothing to do with belief.[37]

His non-cognitivism is of course the reason why Kuhn can, and even must, sentence all present and future philosophers of science to the torments of the damned: that is, to reading the sociology of science. If he were right there would indeed be nothing else for them to do. This prospect inspires in our other authors the horror it should. Yet they could not without inconsistency deny the justice of the sentence.

APPENDIX TO CHAPTER ONE

Helps to Young Authors (I)

Neutralising success words, after the manner of the best authorities

How to rewrite the sentence: Cook discovered Cook Strait.

Lakatos: Cook 'discovered' Cook Strait.

Popper: Among an infinity of equally impossible
alternatives, one hypothesis which has been
especially fruitful in suggesting problems for
further research and critical discussion is the
conjecture (first 'confirmed' by the work of Cook)
that a strait separates northern from southern New
Zealand.

Kuhn: It would of course be a gross anachronism to call
the flat-earth paradigm in geography *mistaken*. It
is simply incommensurable with later paradigms:
as is evident from the fact that, for example,
problems of antipodean geography could not even
be posed under it. Under the Magellanic paradigm,
however, one of the problems posed, and solved
in the negative, was that of whether New Zealand
is a single land mass. That this problem was solved
by *Cook* is, however, a vulgar error of whig
historians, utterly discredited by recent
historiography. Discovery of the Strait would have
been impossible, or at least would not have been

science, but for the presence of the Royal Society on board, in the person of Sir Joseph Banks. Much more research by my graduate students into the current sociology of the geographical profession will be needed, however, before it will be known whether, under present paradigms, the problem of the existence of Cook Strait remains solved, or has become unsolved again, or an un-problem.

Feyerabend: Long before the constipated and boneheaded Cook, whose knowledge of the optics of his telescopes was minimal, rationally imposed, by means of tricks, jokes, and non-sequiturs, the myth of Cook Strait on the 'educated' world, Maori scientists not only 'knew' of the existence of the Strait but often crossed it by turning themselves into birds. Now, however, not only this ability but the very knowledge of the 'existence' of the Strait has been lost forever. This is owing to the malignant influence exercised on education by authoritarian scientists and philosophers, especially the LSE critical rationalists, who have not accepted my criticisms and should be sacked. "No doubt this *financial* criticism of ideas will be more effective than ... intellectual criticism, and it should be used." *(Boston Studies in the Philosophy of Science,* Vol. LVIII, 1978, p. 144.)

CHAPTER TWO

SABOTAGING LOGICAL EXPRESSIONS

I WILL call a statement a "logical" one, or a "statement of logic" if and only if it implies something about what the logical relation is between certain propositions; and the word or phrase, in virtue of which it has this implication, I will call a "logical expression". Thus for example, any substitution-instance of "P entails Q", or of "P is inconsistent with Q", will be a logical statement; and in it "entails" or "is inconsistent with" will be a logical expression. These statements and expressions are even *purely* logical ones; because "P entails Q", or "P is inconsistent with Q", implies nothing about P and Q *except* a statement of what their logical relation is. But a statement or expression can be logical without being purely so. For example, "P is a proof that Q" is a logical statement, because it implies that P entails Q, and in it "is a proof that" is a logical expression; but, they are not purely logical, because "P is a proof that Q", besides implying what the logical relation is between P and Q, implies other things as well, such as that P is true.

A statement of what the logical relation is between P and Q, is equivalent to certain statements about how rationally conclusive certain inferences are: certain inferences, namely, from the truth or the falsity of P to the truth or the falsity of Q. The strongest logical statements such as that P entails Q, or that P is inconsistent with Q, imply that certain of these inferences are *completely* conclusive. The logical relations between P and Q implied by such statements as these are therefore like frictionless pipes along which knowledge can *travel,* and travel without loss. In virtue of P entailing Q, if you know that P you can arrive at knowledge that Q, by travelling along that logical relation; but even if you did not arrive at this knowledge in that way, your claim to know that Q cannot be consistently denied while your claim to know that P is admitted, if P does entail Q. Again, if P is inconsistent with Q then someone who admits your claim to know that P cannot consistently deny your claim to know that Q is false.

Suppose I say that the proposition Q has been proved. Then I have used a success-word. I might then say that in particular it is the truth of P that proves Q, or is the proof (or a proof) that Q. Here I have used a logical expression. Now let us suppose, however, that in my first remark I was neutralising the success-word: using the word "proved" without its implication of truth. Then clearly, to be consistent, I must do something to the logical expression in my second remark, which is like neutralising a success-word. For that remark implies that P entails Q, and I said that P is true; so I will not be able to go on doing what I began by doing, avoiding the implication that Q is true, if I leave intact that implication of my logical statement.

Similarly with "refuted" and "refutation", for example. Suppose I first say that Q has been refuted, and then say that it is the truth of P which refutes Q, or is the refutation (or a refutation) of Q. In the first remark "refuted" is a failure-word, implying the falsity of Q; in the second, the cognate words are logical ones, since in virtue of them my remark implies that P is inconsistent with Q. But now suppose that in my first remark I was neutralising the failure-word, taking out its implication of falsity. Then, to be consistent, I clearly must do something, to the logical expression in my second remark, which is like neutralising a failure-word. Having admitted the truth of P, I will not be able to continue, as I began, avoiding the implication that Q is false, if I allow my logical statement to retain the implication that P is inconsistent with Q.

What consistency requires in such cases is *like* neutralising a success- or a failure-word. For I must *use* a logical expression, and *appear* by doing so to make a statement of logic, that is, appear to imply something about the logical relation between propositions; but at the same time, I must really not do so.

The process in question cannot *be* that of neutralising a success- or a failure-word, however. Some logical words *are* indeed success-words as well. For example, "(is) proved" and "proof" are such; just because the connection in meaning is so close between saying that Q is proved, and saying that there is some truth P which is a or the proof of Q. Similarly "(is) refuted" is a logical word as well as being a failure-word, and "refutation" is both too. But on the other hand, success- or failure-words need not be logical words. "Knowledge", for example, is not. That Q is known, implies nothing about the logical relation of Q to any other proposition. And contrariwise, logical words need not be success- or failure-words. For example, purely logical expressions like "entails" or "is inconsistent with" only imply something about logical relations, nothing about cognitive success or failure; and not having any such implication, they cannot be deprived of it.

Nevertheless, we recall, the logical relation between two propositions, which is what a logical statement implies something about, may be a path along which cognitive achievement can travel; and even travel without loss in the case of logical relations like entailment or inconsistency. The practice we are now thinking of, of using logical expressions so as to appear to make a statement of logic, but without in fact implying anything about logical relations, is one which would make such travel impossible. Neutralising a success-word is a device for wiping out cognitive achievement after it has arrived. Its counterpart for logical expressions would prevent cognitive achievement, if it had to travel along logical relations, which almost all cognitive achievement has to do sooner or later, from ever arriving. It is like blowing up railway tracks, holing water pipes, or cutting power-lines. Let us call it "sabotaging" a logical expression.

This is the second of the two main literary devices by which our authors make irrationalist philosophy of science plausible. The first, the use of success-words (though neutralised), is of course a device which makes *directly* for plausibility. Sabotaging logical expressions does not do this, but it is an essential auxiliary to the first device. A writer who often took the implication of truth out of "proved", but never took the implication of *entailment* out of "proof", or who often took the implication of falsity out of "refuted" but never took the implication of *inconsistency* out of "refutation", would be in a position hopelessly exposed to criticism. Our authors have not been so careless.

This will be proved by examples later. Obviously I cannot prove that philosophers of science who are not irrationalist do not sabotage logical expressions as often as our authors do. But I think that my examples will be found sufficiently distinctive of the kind of English our authors write. And anyone who tries to match these examples with examples of the sabotage of logical expressions drawn from the writings of Hempel or Carnap, say, will find that experiment instructive.

But sabotaging logical expressions is not only a device

which is for our authors an essential auxiliary to that of neutralising success-words. It has a most important and distinctive effect of its own, directly on the literary fabric of their philosophy of science, and indirectly in giving that philosophy plausibility. For it generates what I call "ghost-logical statements". To explain what I mean, I need to anticipate slightly.

One way to sabotage a logical expression, and the way which is most common in our authors, is to embed a logical statement in a context which can be broadly described as epistemic. A schematic example, and one not likely to occur in our authors, is this: instead of saying "P entails Q", which is of course a logical statement, to say "P entails Q according to most logicians, ancient, medieval, and modern".

The latter statement, unlike the former, is not a logical statement at all: it implies nothing about the logical relation of P to Q. It is really just a statement, contingently true or false, about the *history* of logic. Yet at the same time it makes the strongest possible suggestion, not only that a statement of logic *is* being made, but that one is being made from which no rational person will dissent. The context "according to most logicians (etc.)" sabotages the logical expression "entails"; yet suggestions of logic are so artfully blended with implications of history that the statement is a kind of mirage of a logical statement being made: it is a ghost-logical statement.

Now ghost-logical statements have, while logical statements lack, a characteristic of the utmost importance (and ghastliness): they are *absolutely immune to criticism on logical grounds*. For consider "P entails Q according to most logicians (etc,)." Suppose some one attempts to criticise it on logical grounds, and suppose that the outcome of his attempt is the most favourable possible for the critic: that is, he succeeds in showing that after all P does *not* entail Q. What is that to the purpose?

Nothing. For the statement he set out to criticise never did imply that P does entail Q.

That statement is also, we see, *virtually* immune to criticism on *historical* grounds too. The task of historical criticism of it would be at once so enormous, and so indefinite that, if a critic did set out on that venture, you could rely on his never returning from it. And this virtual immunity even to historical criticism is possessed by very many ghost-logical statements (though of course not by all).

Confronted with our ghost-logical statement, however, a potential critic is not likely to be able to contemplate distinctly either the possibility of logical criticism or the possibility of historical criticism of it. What is far more likely is that his critical powers will be paralysed, and he will not know *how* to react to the given statement: for the ghost-logical statements produced by epistemic embedding are typically not only immune to criticism, but actively paralyse it. And the reason is clear. Such a statement is like a statue of Janus, forever pointing the potential critic in opposite directions at once: implying that historical criticism (even if practically impossible) would be relevant, and logical criticism not, while at the same time irresistibly suggesting that logical criticism would be relevant, and historical not.

When our authors use this method of sabotaging a logical expression, their epistemic contexts will be drawn of course from the history of science, not from the history of logic as in the above example. But the result will be the same, namely a ghost-logical statement. And in fact ghost-logical statements are common enough in our authors (and their followers), and peculiar enough to them, to constitute, along with neutralised success-words, a literary hall-mark by which any writing of theirs can be identified as such.

While the neutralising of success-words contributes directly to the plausibility of irrationalist philosophy of science, then, the sabotage of logical expressions by epistemic embedding does not. But it indirectly makes an enormous contribution of its own to the plausibility of our authors: it enables them to make statements, about the relations between the propositions of science, which appear to be statements of logic, and yet which possess absolute immunity to logical criticism. For it generates not logical but ghost-logical statements; and these (as well as being in most cases virtually immune to historical criticism) are always and absolutely immune to logical criticism. How great an advantage in philosophy such immunity is, and how important as an indirect aid to plausibility, no philosopher will need to be told.

L OGICAL expressions, whether purely logical or not, can be divided into strong and weak ones, just as success-words can; and as I have said, some of them are success-words as well. Examples of strong logical expressions are "entails", "proves", "verifies", "has as a special case". Weak ones include "is consistent with", "supports", "confirms", "is a special case or example of". Logical expressions weaker still include "explains" and "solves the problem of". (Here as elsewhere I generally take verbs as my logical expressions, but of course the cognate noun will be a logical expression too.)

All the above are in an intuitive sense *positive* logical expressions. Corresponding to them, and for the most part trivially intertranslatable with them, are negative logical expressions. Strong negative ones include "is inconsistent with", "disproves", "refutes", "falsifies", "is a counter-example to", "clashes with". Weak negative expressions include "disconfirms", "is an anomaly for", "poses a problem for", "cannot explain", "fails to predict".

In our authors, the sub-class of logical expressions which

is most prominent is that of the strong negative ones. The historical reasons for this are obvious. Popper had undertaken, in *The Logic of Scientific Discovery,* to display the entire logic of science without departing once from the vocabulary of deductive logic; and even the positive part of that vocabulary was not needed, he thought, except for the uncontroversial work of describing the 'downward' articulation of laws and theories. For the rest, Popper claimed, everything in science could be understood with the aid of "falsification", "refutation", or other expressions implying inconsistency; that is, of the strong negative logical expressions. Our other authors are of course in this respect very much under Popper's influence. It was therefore to be expected, and it is in fact the case, that when any of our authors sabotages a logical expression, it is almost always a strong negative one.

If our authors are willing, as I will prove by examples that they are, to sabotage even strong negative logical expressions, they will be still more willing to sabotage weak ones, negative or positive. And if they sabotage strong negative ones, then this practice will confer on what they write a wider immunity to logical criticism than the same thing would do for other writers of a less 'deductivist' tendency: just because our authors make such comparatively little use of logical expressions of any other kind.

WHAT ways are there, then, of sabotaging logical expressions? That is, of seeming to imply something about the logical relation between propositions, without actually doing so.

First, it may be done by enclosing a logical expression in quotation-marks. "P 'entails' Q" can be used in such a way, or in such a context, as to suggest, what of course it does not imply, that P entails Q.

This is a tiresome subject, but it cannot be entirely omitted. Popper, and any philosopher of science much influenced by Popper, will sometimes be found sabotaging by quotation-marks some of the weak positive logical expressions at least: for example "confirms". And Lakatos in particular is forever using quotation-marks to sabotage even the strongest logical words, such as "proof" and "refutation": see his *Proofs and Refutations; passim*. Most of what was said in Chapter One about Lakatos and quotation-marks could in fact be repeated here, because it happens that most of his victims are logical words as well as being success- or failure-words. But I will add just one short example.

"If a theory is refuted, it is not necessarily false. If God refutes a theory, it is 'truly refuted'; if a man refutes a theory, it is not necessarily 'truly refuted'."[1]

This example is the more sad, because it seems clear that what Lakatos tried to say was that if God, unlike man, refutes a theory, then it is truly refuted. If so, however, his tic was too strong for him, and he could not say it. For as the sentence came out, we see, even God's refutations were sabotaged by quotation-marks.

Let us leave this depressing topic.

A second way to sabotage a logical expression, and the way our authors use most, is the one I have already mentioned in anticipation: *by embedding a statement of logic in an epistemic context.*

Of course such embedding need not result in sabotage of the logical expression. It need not even cut off the embedded statement's implication about logical relations. "Everyone knows that P entails Q" is, among other things, a logical statement in my sense, not a ghost-logical one: it implies that P entails Q, just because of the success-word "knows" in the context. But even if the implication about logical relations *is* cut off by embedding, there need not result any sabotage of the logical expression. "Some people think that P entails Q", for example, is not logical, since it implies nothing about the logical relation of P and Q; but it is not ghost-logical either, since there is nothing in it to suggest that it *does* have such an implication. The statement is a plain historical one, and does not pretend to be anything else.

But consider the following schematic examples, in which a logical expression is sabotaged by epistemic embedding. "Any scientist would regard P as entailing Q." "P entails Q on the Copenhagen interpretation." "Given the conceptual scheme of special relativity, P entails Q." "Once R was discovered, though not before, P could be seen as entailing Q." "P does indeed entail Q, once Galileo's paradigm is adopted." "Q was a logical consequence which could hardly be overlooked once P was added to the hard core R of the research programme... A scientist who accepted P but rejected Q would be regarded by his profession as violating one of its most basic values, consistency."

It is rather easy, isn't it, to sabotage logical expressions by epistemic embedding? For it is being done in these examples fairly effectively, and even with an approach in some cases to our authors' individual *styles* of sabotage. (You do not need to be an anti-saboteur specialist to see, in the ruins of a logical relation in the last example, a fair imitation of the handiwork of Kuhn.) Yet it is being done here under the most unfavourable possible conditions. My 'propositions' being mere dummies, "P entails Q" could not help, by its own truth or plausibility, to second the suggestion (the false suggestion) that a statement of logic is being made. The main logical expression used here, "entails", is one of those least easy to sabotage. A weak logical expression such as "confirms" succumbs far more readily to sabotage; as may be seen by the fact that the first example, say, will be still more easily mistaken for a logical statement if we replace "entailing" by "confirming". My epistemic beds were of necessity imaginary, and above all too *short* to be very lifelike; quite unlike those vast beds of detail, drawn from the actual history of science, which are available to Lakatos, Kuhn and Feyerabend, for suffocating logical expressions. Yet even under all these handicaps, we see, it is not at all hard to set up a suggestion, and a suggestion of almost any degree of strength that might be desired, that P entails Q; even though, because of the epistemic embedding, one has actually implied nothing of the kind. If you were a writer likely in any case to switch from the logic to the history of science and back again, and still more if you considered yourself licensed to do so, and as fast and as often as you like, you could positively leave this trick to work itself.

In some of the above examples there is a hint of another vice as well, something quite additional to the sabotaging of a logical expression. This is what I have elsewhere called "misconditionalisation":[2] that is, for example, saying that if R, then P entails Q, when what you really mean is that the conjunction of R and P entails Q. Misconditionalisation, however, is a vicious process performed on logical statements. (It will often turn true ones into false.) Sabotage of a logical expression, on the other hand, is a literary device for appearing to make a logical statement, without actually doing so. Misconditionalisation can be used to *assist* the sabotage of logical expressions by epistemic embedding, and in our authors it is in fact sometimes so used. But that is as far as the connection between the two extends.

The examples given above, as well as being schematic, were of the simplest type possible; and this in two respects.

First, the logical relation which was sabotaged was one between particular propositions: between some concrete values of the propositional dummies P and Q. But of course it is equally possible, and for a philosopher it is often more natural, to sabotage the logical relation between P and Q where these are *kinds* of propositions: where P stands for theories, say, or law-statements, and Q for, say, observation statements, or again, statements of initial conditions. In other words, the logical statement which is embedded in a ghost-logical statement need not be singular, but may be general, and of any degree of generality. Example: "A scientist would never regard a law-statement as entailing any statement of initial conditions." And in our authors, sabotage by epistemic embedding is in fact more usually of general logical statements than of singular ones.

Second, in the above examples there was no *iteration* of epistemic embedding; but there easily can be such a thing. A logical statement can be sabotaged, by being embedded in an epistemic context, and then this whole thing, the ghost-logical statement, can in turn be embedded in another epistemic context. Schematic example: "Logicians generally assume that any scientist would regard P as inconsistent with Q". There is no theoretical limit, of course, to how far this 'layering' of epistemic contexts, over an original statement of logic, may go. In our authors there is hardly any practical limit either. Moreover the ghost-logical effect of the first embedding, that is, the false suggestion of a logical statement being made, may pass undiminished through the second embedding, or may even be amplified by it. It will depend on the nature of the second epistemic context whether this happens or not. Here is an example in which iteration does amplify, or at least does not diminish, the ghost-logical effect of the first embedding. "Most philosophers of science, since reading Popper, Kuhn, Lakatos, and Feyerabend, have agreed that scientists never regard a theory or a law-statement as falsified by a single observation-statement."

This example is not one which could occur in our authors themselves, of course, though there are plenty of others that do. But in the writings of their followers (those who are doing 'normal science', as it were, in the wake of the irrationalist revolution), instances of such iterated epistemic embedding are especially common. And one can easily see why: they make assurance doubly sure. The first epistemic context, "scientists never regard", is sufficient on its own, as we have seen, to confer on what is said total immunity to logical criticism. The second epistemic context, "philosophers of science ... have agreed", while it amplifies if anything the false suggestion that a statement of logic is being made, at the same time buries the embedded logical statement so deep in sociology, that omnipotence itself might despair of ever dredging it up again. As for ordinary philosophers, as Lakatos calls them (actually of course he calls them "'ordinary' philosophers"[3]), who might be tempted to criticise this remark, let them consider the utter hopelessness of that undertaking.

Historical criticism, if that is what is aimed at, would need to begin with the outer context, about philosophers. Should the critic's work there be ever done, he still has the inner context, about scientists, before him. How can he succeed here? Being an ordinary philosopher he probably does not know enough about the history of science; and if he does, then he also knows that any actual episode in the history of science is so complicated that he will never be able to put it beyond dispute that in it a scientist regarded (say) a law-statement as falsified by a single observation statement. As for logical criticism of the remark, to which the philosophical critic is more likely to be drawn, it must for a start be long: much longer, at least, than the remark which he is criticising. Again, it will require much tedious insistence on obviously true statements of logic: statements about the relation between propositions (those mummified objects of the ordinary philosopher's art). It must therefore be boring. But let his logical criticism be ever so good: let him prove to perfection that an observation-statement can be inconsistent with, and therefore can falsify, a law-statement. He has still only wasted his own and others' time, in proving an irrelevance. For it never was said that an observation-statement cannot falsify a law statement. Only that most recent philosophers of science have *agreed* that scientists never *regard* them as doing so. And that is a proposition which is so very far from being a statement of logic, a contribution (whether a true or a false one) to the logic of science, that it actually belongs to the sociology of the philosophy of science!

Such are the joys in store for anyone who would attempt to criticise such a representative expression of irrationalist philosophy of science as we have just been considering. And such are the advantages, correspondingly, to irrationalism, of the sabotage of logical expressions by embedding them in an epistemic context or in more than one. But while this device bestows, on those who are willing to use it, virtual immunity to all criticism, and absolute immunity to logical criticism, I entertain some hopes that it may not be entirely proof against simple exposure, as the deceitful literary device it is. Let us turn to some specific and representative passages of our authors.

THE use of epistemic embedding to sabotage a logical expression is less common in Popper than in any of our other authors. Even when he does it, he does it in a more diffuse way than they usually do, and (what may be connected with that) not with quite the same unclouded conscience. Yet, characteristically, it was he who began the practice, and by the authority of his example gave it currency.

His most influential act of sabotage occurs in a part of *The Logic of Scientific Discovery* which is seldom read, or at any rate remembered, by any but adepts. The instance must have been sufficiently grievous, because even people not otherwise apt to criticise Popper complained of it,[4] and what is more remarkable still, Popper himself later said in print that what he had written at this place was "not to my own full satisfaction".[5] To readers in whom the critical faculty is not entirely extinct, the episode has afforded a certain amount of hilarity. To our other authors, by contrast, what it afforded was a model and a licence for their own efforts in the way of sabotaging logical expressions. If what Popper did here was not to his own full satisfaction, it certainly was to theirs.

The propositions in question were unrestricted statements of factual probability: that is, contingent unrestricted propositions of the form "The probability of an F being G is $= r$", where $0 < r < 1$. For example, H: "The probability of a human birth being male is $= .9$". Concerning such propositions Popper had fairly painted himself into a corner. For he had maintained (1) that some such propositions are scientific; (2) that none of them is falsifiable (i.e. inconsistent with some observation-statement); while he had also maintained (3) that only falsifiable propositions are scientific. (The reason why (2) is true is, of course, that H is consistent even with, for example, the observation-statement E: "The observed relative frequency of males among births in human history so far is $= .51$".)

Popper draws attention with admirable explicitness[6] to this—to put it mildly—*contretemps*. He puts it almost equally mildly himself, however. For he insists on calling the conjunction of (1), (2) and (3) a "problem" ("the problem of the decidability"[7] of propositions like H); when in fact, of course, it is a contradiction. The reader can hardly fail to be reminded of Hume's complaint about the absurdity of the "custom of calling a *difficulty* what pretends to be.a *demonstration* and endeavouring by that means to elude its force and evidence."[8] But Popper's 'solution' to his problem was far more remarkable than even his description of it, and indeed was of breathtaking originality.

It consists—or I should say, it appears to consist, because there is another interpretation of Popper possible here, though one which makes his solution far less satisfactory still, which will be discussed later—in making frequent references to what it is that scientists do when they find by experience that s, the observed relative frequency of G among F's, is very different from r, the hypothesised value of the probability of an F being G. What scientists do in such circumstances, Popper says, is to act on a methodological convention to neglect extreme improbabilities (such as the joint truth of E and H); on a "methodological rule or decision to regard ... [a high] negative degree of corroboration as a falsification"[9], that is, to regard E as falsifying H.

Well, no doubt they do. But obviously, as a solution to Popper's problem, this is of that kind for which old-fashioned boys' weeklies were once famous: "With one bound Jack was free!" What will it profit a man, if he has caught himself in a flat contradiction, to tell us about something that scientists do, or about something non-scientists don't do, or anything of that sort? To a logical problem such as the inconsistency of (1), (2) and (3) there is of course—can it really be necessary to say this?—no solution, except solutions which begin with an admission that at least one of the three is false. But least of all can there be any *sociological solution*.

For our purposes, however, what is important about the episode is the following. The pairs of propositions we are talking about are pairs such as E and H. As (2) implies, and as is in any case obvious, E is consistent with H. But the logical word "falsifies" or its cognates, applied to a pair of propositions, implies that their logical relation is that of inconsistency. So to say that E falsifies H would be to make a logical statement which is false, necessarily false, and obviously false. So Popper will not say *that*. What he says instead are things which, however irrelevant to his problem, are at least true (even if only contingently true). Such as the following. That "a physicist is usually quite well able to decide" when to consider an hypothesis such as H "'practically falsified'"[10] (namely, when he finds by experience, for example, that E). That "the physicist knows well enough when to regard a probability assumption as falsified"[11] (for example he will regard H as falsified by E). That propositions such as H "in empirical science ... are used as falsifiable statements."[12] That given such an observation-statement as E, "we shall no doubt abandon our estimate [of probability, that is, H] in practice and regard it as falsified."[13]

These are very models of how to sabotage a logical expression by epistemic embedding, or of ghost-logical statements. They use a logical expression, one implying inconsistency, but they do not imply the inconsistency of any propositions at all. They are simply contingent truths about scientists. Yet at the same time there is a suggestion that not only *is* a logical statement, implying inconsistency, being made, but that one is being made with which no rational person would disagree. This suggestion is in fact so strong as to be nearly irresistible, and it comes from several sources.

First, Popper's references to a rule, decision, or convention, imply that when scientists regard E as falsifying H, they cannot be *wrong*: and they therefore serve to suggest that they are right. Second, there is the fact that *scientists* regard E as falsifying H, and that they are unanimous in doing so. How can a reader suppose that scientists, all scientists, are *mistaken* in regarding E as inconsistent with H? He might almost as easily suppose all philosophers mistaken in regarding a Barbara syllogism as valid. Third, and most important of all: the reader's own common sense - and it is his *logical* common sense - emphatically seconds the statement of logic which here appears, by *suggestio falsi,* to be being made. He knows, as everyone (near enough) knows, that given E, it *is* rational to infer that H is false. And since scientists, as these statements report them, seem to be saying only very much that same thing, the reader is disposed to think that the scientists are right. And if they *are* right, it is clearly a point of *logic* on which they are right.

The suggestion, coming from all these sources, that a logical statement, and a true one, is being made, is so strong, in fact, that to many people it will appear perverse, or at least pedantic, to resist it. What is there, then, to object to, in the statement that scientists regard E as falsifying H?

Simply that its suggestion, that a statement of logic is being made, is false; and that *suggestio falsi* is not better, but worse, the stronger the suggestion is. The statement is only a ghost logical statement. It implies nothing whatever about the logical relation between E and H. A logical word, "falsifying", is used indeed, but its implication of inconsistency is sabotaged by the epistemic context about scientists. This is cold-blooded murder of a perfectly good logical expression, in exchange for a handful of sociological silver about scientists.

What makes the case more unforgivable is that the logical expression here sabotaged is not only a strong or deductive-logical expression, but the one which is, of all deductive-logical words, Popper's own particular favourite; and that he had just a few pages before undertaken that, however others might succumb to non-deductive logic, he never would, but that in *his* philosophy all relations between propositions of science would be "fully analysed in terms of the classical logical relations of *deducibility and contradiction*"![14]

I mentioned earlier that all our authors are of a marked deductivist tendency, and that therefore, when they can no longer avoid a weak or non-deductive logical expression such as "confirms", they will sometimes sabotage it by quotation-marks. We have now seen an example of another strategy that our authors use, when non-deductive logic threatens to break into their philosophy. This is, to retain the deductive-logical words, but deprive them of their deductive-logical meaning, by embedding them in epistemic contexts about scientists. A painful spectacle this: like the citizens of a besieged town, when the besiegers are on the point of breaking in, strangling their own children.

We will see later that our other authors repeat on their own behalf, and extend, what Popper did to the logical relation between statements of factual probability and observation-statements. For the present, let us turn to another, and a less special case: the relation between scientific theories and other statements. Here it is possible to display a series of statements, beginning with Popper and extending through our other authors, in each of which "falsifying" or some equivalent logical expression is sabotaged by epistemic embedding. The members of this series are not only linked by strong family resemblance: it is reasonable to believe that, as a matter of history, the other members of the series grew out of the first one.

The better to display both the nature of these statements and the continuity between the members of the series, I will first give a version of my own which will be a composite-photograph, as it were, of many things actually said by one or more of our authors. No one familiar with our authors will dispute the verisimilitude of my versions. But afterwards I will give, for each statement in this list, at least one of the actual passages which have gone to make up the composite version.

(i) 'Only a low-level corroborated theory will be accepted as falsifying a scientific theory.'

(This is the thesis of Popper which is, I believe, the germ of the following three others.)

(ii) 'In the actual history of science, as distinct from the distortions of it by philosophers, theories are never regarded as refuted by reports of observations, or experiments: a scientific theory can be defeated only by another theory.'

(This thesis, although in fact Popperian enough, is more usually associated with our other authors, and is sometimes even advanced by them as a criticism of Popper.)

(iii) 'It is often only with the hindsight provided by a later and rival theory T^1 that a certain experiment is seen as "crucial" against, that is, as a falsification of, an earlier theory T.'

(This thesis is Lakatosian, of course, but is also entirely congenial to Kuhn and Feyerabend.)

(iv) "The anomalies which beset an earlier paradigm are considered intolerable, that is, as *logically* compelling its abandonment, only after the shift to a new paradigm has been made, and even then only by those scientists who have made the shift."

(This is Kuhnian, of course, but again entirely congenial to Lakatos and Feyerabend.)

The following are some of the actual passages from which the foregoing composites have been made.

In connection with (i): "... non-reproducible single occurrences are of no significance to science. Thus a few stray basic statements contradicting a theory will scarcely induce us to reject it as falsified. We shall take it as falsified only if we discover a *reproducible effect* which refutes the theory. In other words, we only accept the falsification if a low-level hypothesis which describes such an effect is proposed and corroborated."[15]

In connection with (ii): "... a clash [with observation] may present a problem (major or minor) [for a theory], but in no circumstance a 'victory' [for observation]. Nature may shout *no,* but human ingenuity—contrary to Weyl and Popper—may always be able to shout louder. With sufficient resourcefulness and some luck, any theory can be defended 'progressively' for a long time, even if it is false." And "... a rival theory, which acts as an *external* catalyst for the Popperian falsification of a theory, here becomes [i.e. on Lakatos's methodology] an *internal* factor."[16]

In connection with (iii): "The anomalous behaviour of Mercury's perihelion was known for decades as one of the many yet unsolved difficulties in Newton's programme; but only the fact that Einstein's theory explained it better transformed a dull anomaly into a brilliant 'refutation' of Newton's research programme. Young claimed that his double-slit experiment of 1802 was a crucial experiment between the corpuscular and the wave programmes of optics; but his claim was only acknowledged much later, after Fresnel developed the wave programme much further 'progressively' and it became clear that the Newtonians could not match its heuristic power. The anomaly, which had been known for decades, received the honorific title of refutation, the experiment the honorific title of 'crucial experiment', only after a long period of uneven development of the two rival programmes."[17]

In connection with (iv): "Ordinarily, it is only much later, after the new paradigm has been developed, accepted, and exploited, that apparently decisive arguments [against the old paradigm] are developed. Producing them is part of normal science, and their role is not in paradigm debate but in postrevolutionary texts."[18] Again: "Though the historian can always find men - Priestley for instance - who were unreasonable to resist [a new paradigm] for as long as they did, he will not find a point at which resistance becomes illogical or unscientific."[19]

STATEMENTS such as I have here either quoted or paraphrased will be admitted to be representative of our authors' writings. They even embody a considerable amount of the substance of their philosophy of science, and they certainly exemplify a very characteristic way our authors have of expressing that philosophy. They are a fair sample of our authors' contributions to that enterprise on which they are all engaged, and their contributions to which constitute their main claim on their readers' attention: the enterprise of making known the logic of science.

It is therefore worthwhile to point out that not one of them is a logical statement at all. Not one of them has a single implication about the logical relation between any propositions whatever. They all indeed use logical expressions, and strong ones: "falsifying", "refuting", "decisive argument against", or some cognate equivalent. But every time such an expression occurs, it is sabotaged by being embedded in an epistemic context about scientists: about how scientists 'regard', 'consider', 'take', 'see', etc., the relation between certain positions. In short, they are one and all ghost-logical statements, and nothing more.

Enough examples, and sufficiently representative ones, have perhaps now been given, to enable the reader to begin to realise how extremely common in our authors is the sabotage of logical expressions by epistemic embedding. Once you mix the history with the logic of science, the possibilities of such sabotage are limitless; and almost every possibility has been realised. Recall for example Kuhn's willingness to dissolve even the strongest logical expressions into sociology about what scientists *regard as* decisive arguments; recall that the logical expressions most important to him (namely the positive "solves the problem of", and the negative "is an anomaly for") are weak ones, and are therefore easily sabotaged; recall his express and repeated assertion that what *constitutes* solution of a problem is paradigm-relative; and you will see that his entire philosophy of science is actually an engine for the mass-destruction of all logical expressions whatever: a 'final solution' to the problem of the logic of science. Then there is that variety of iterated epistemic embedding which is an especial pest in Lakatos: 'According to sophisticated theoretical conventionalism' (or whatever) 'scientists never regard such-and-such as inconsistent with so-and-so', etc., etc. But these are the blockbusters of sabotage. The small arms fire, which almost never stops, is better represented by the following quotation from Popper, in which he is discussing positive scientific knowledge. Now "positive scientific knowledge" may perhaps be not obviously a logical expression, but a logical expression it must be. "Positive scientific knowledge" must at least entail "well-confirmed theories", for example; and here is what Popper says. "In my view, all that can possibly be 'positive' in our scientific knowledge is positive *only* in so far as certain theories are, at a certain moment of time, preferred to others in the light of our critical discussion, which consists of attempted refutations ...".[20]

Thus, where the reader expects, and non-irrationalist philosophy would require, the word "preferable", what Popper actually says is "preferred"; and so the logical expression "positive scientific knowledge" is quietly sabotaged, by a context referring only to the actual theory-preferences of people (presumably scientists). There are literally hundreds of sentences in Popper like this. He is about as sensitive to the difference between an evaluative word (like "preferable") and a descriptive one (like "preferred") as Mill in a famous passage[21] showed himself to be to the difference between "desirable" and "desired".

Such examples also enable us to understand something which is especially prominent in Kuhn, and which is otherwise baffling: what might be called his 'tautological optimism' about science. See his [1970a] Chapter XIII *passim,* and for example the following. "If I sometimes say that any choice made by scientists on the basis of their past experience and in conformity with their traditional values is *ipso facto* valid science for its time, *I am only underscoring a tautology*."[22]

The reader is dumbfounded: the *validity* of current science guaranteed by ordinary scientific behaviour plus tautology? There has to be a catch in that! How could Kuhn come to write such a thing? Well, we now know how. If, like Kuhn, you cannot tell the difference between talk about the logical relation of current evidence P to current theory Q, and ghost-logical talk about what scientists *consider to be* the logical relation of P to Q, why, then, the connection between valid science, and what scientists currently consider to be so, will of course seem to you to be perfectly analytic.

It may be worthwhile to add another instance in which it is a weak positive logical expression, rather than a strong negative one, which is sabotaged. (This example is a composite-photograph one, but will be easily recognised as representative.) "Scientists consider a theory T^1 better confirmed than a theory T, if T^1 explains all that T does, avoids the failures of T, and predicts facts which T does not." This kind of example is interesting for two reasons. One is that, because the embedded logical statement ("T^1 is better confirmed (etc.)") is so highly plausible, the sabotage of it by "Scientists consider" is, to a correspondingly high degree, inconspicuous: indeed, it is almost imperceptible. The other reason is, that such ghost-logical statements as this one have a special advantage for *deductivist* authors. For the embedded logical statement is, of course, one belonging to non-deductive logic, and therefore it would, if asserted naked, entangle the author in non-deductive logic; whereas when it is clothed in an epistemic context about scientists, it leaves one's options open and one's deductivist record unblemished. There is a slight drawback, of course, that one has now not made a logical statement at all! But then, no one is likely to notice that: at least, they never have.

I have not implied, and it is not true, that our authors are the *only* ones in recent philosophy of science who ever sabotage a logical expression by epistemic embedding. In fact, easily the most influential ghost-logical statement of the century is one which is not usually associated with them at all; though they find it congenial, and, as we might have expected, Popper actually anticipated it. This is 'the Quine-Duhem thesis': that "any statement can be held true come what may, if we make drastic enough adjustments elsewhere in the system... Conversely, ...no statement is immune to revision."[23]

Would-be critics of this thesis have been mystified by *its* immunity to revision. They would not have been, if they had come to it from a course of reading in our authors, as we do now. For our ears are by now accustomed to detect the fatal premonitory sounds of sabotage—'regarded as', 'accepted as', etc., prefixed to a logical expression—and once we hear, in Quine's word "held", the doom of a logical expression again pronounced, we *expect* immunity to logical criticism to ensue. The thesis means, of course, that any scientific theory can be held to be consistent with any observation-statement, however 'recalcitrant', provided we make drastic enough adjustments elsewhere in the system. This thesis is undoubtedly true. But what kind of truth is it?

It seems to be a statement of logic. For it seems to imply that a certain logical relation, namely consistency, exists between certain kinds of propositions. But look more closely: the thesis does not imply that observation-statements are always consistent with scientific theories. It does not imply that they *ever* are. It does not *imply anything* about the logical relation between any propositions whatever.

In fact the thesis is simply the most trivial of contingent truths about human beings: that given any proposition whatever, a scientist (or anyone) can take it into his head to affirm it, and can then stick to it through thick and thin. *Of course* it is true that "any statement can be held true come what may, *if we make drastic enough adjustments elsewhere in the system*": for the simple reason that any statement can be held true come what may, *with or without* making 'adjustments elsewhere'. Quine's proviso then, (here italicised), was entirely unnecessary for the truth of what he said. Its only function was that of *suggestio falsi*: to generate the illusion that a statement of logic, and in particular one which implies that certain propositions are consistent with one another, was being made.

A THIRD way to sabotage a logical expression is to embed it in a context which is not epistemic, but of a kind I will call "volitional". This kind of context makes the logical relation, implied by the logical statement embedded in it, an object not of anything epistemic (such as belief), but of the will. The logical relation between propositions is now spoken of as being, by some one or other, decided or chosen or made to be, entailment, inconsistency, or whatever. Schematic examples: "Logicians, let us make the Barbara syllogism valid"; "I permit P to be consistent with Q"; "I propose a rule making P and Q inconsistent"; "I propose the adoption of a convention to regard P as entailing Q".

This may not seem a very intelligible way to speak. Still, our authors often do speak in essentially this way. And historically this practice too stems, just as epistemic embedding does, from the *locus classicus* in Popper already discussed, concerning the unfalsifiability of unrestricted statements of factual probability. So we must return to that.

Our example, it will be remembered, of the propositions here in question, was H, "The probability of a human birth being male is = .9". Its unfalsifiability (that is, its consistency with every observation-statement) we exemplified by its consistency even with E, "The observed relative frequency of male births in human history so far is = .51". And Popper's problem, we recall, was that he had asserted (1) that some unrestricted statements of factual probability are scientific; (2) that none of them is falsifiable; and (3) that only falsifiable statements are scientific.

Now I took Popper's solution, when I wrote about it above, to be an historical *report,* by Popper, that scientists do in fact act on a convention to regard E as falsifying H. But what he says may also be interpreted in another way, either instead of this, or (as I believe it should be) in addition to it: as *Popper himself proposing* such a convention.

Lakatos and Kuhn both interpret Popper exclusively in this second way. And some remarks of Popper himself, though made much later, give additional weight to this interpretation. (See the text to footnotes 30 and 31 below.)

On my first interpretation, Popper's sabotage of the logical expression "falsifying" was by embedding it in an epistemic context (about scientists). On the second interpretation, it was by embedding it in a volitional context. For we are now to understand him as *permitting us* to regard E as falsifying H, or as introducing a rule which *makes* E inconsistent with H.

Popper's 'solution' to his problem is, of course, even more amazing on this interpretation than it was on the first. It is bad enough to suggest that you can get yourself out of the contradiction constituted by (1), (2) and (3), by reporting some fact about scientists. But to suggest that you can get out of it by some exercise of your will—by permitting something or proposing something—is more breath-taking still. It is difficult even to understand such a suggestion. Nevertheless, that does appear to be Popper's main suggestion, and we must make the best we can of it.

To Lakatos and Kuhn, at any rate, it presented no difficulty at all. Far from that, they willingly endorse it, and heartily repeat on their own behalf Popper's exemplary act of sabotage. Lakatos writes: "… no result of statistical sampling is ever inconsistent with a statistical theory unless we *make them* inconsistent with the help of Popperian rejection rules …"[24] Again, he writes: "… probabilistic theories … although they are not falsifiable … can easily be made 'falsifiable' by [a] *decision* which the scientist can make by specifying certain rejection rules which may make statistically interpreted evidence 'inconsistent' with the probabilistic theory."[25] Kuhn similarly writes: "… dealing with a probabilistic theory [scientists] must *decide* on a probability threshold below which statistical evidence will be held ' "inconsistent" ' with that theory."[26]

The last quotation is well worth the attention of the connoisseur. The logical word "inconsistent" is first sabotaged by quotation-marks, and not by one but by *two* sets of them. And as though this might still not be quite enough to ensure that the word "inconsistent" no longer means inconsistency, it is sabotaged as well by the volitional context "[scientists] … decide …". Finally, Kuhn's word "must"—but this is best left for an advanced course in the anti-saboteur college.

In Popper, the logical relation between observation-statements and statements of factual probability is the only one sabotaged by embedding in volitional contexts. It is not so in Lakatos or Kuhn. On the contrary, Lakatos speaks quite generally of "making" propositions "unfalsifiable by *fiat*"[27]; that is, as though it were possible to do this to any proposition whatever. Equally generally, Kuhn writes, taking up this phrase of Lakatos, that "Scientists must *decide* which statements to make 'unfalsifiable by *fiat*' and which not."[28]

In these places in our authors, then, it is implied that in at least some cases the logical relation between propositions can be made or chosen by, or be in some way subject to, the will. At many other places in their writings the same thing is suggested. For example, when Feyerabend pleads for majority-rule in science.[29] Or again when Lakatos speaks of the scientist's need, when he has deduced a false conclusion from a complex set of premises, to 'decide where to direct the arrow of *modus tollens*'.[30] The will in question, then, may be that of Popper, or of scientists, or of the majority of people. But all our authors imply, by embedding logical expressions in volitional contexts, that logical relations can be subject to some will.

The major difficulty is simply that of understanding how this could be true. When our authors sabotage logical expressions by embedding them in *epistemic* contexts, the result is a ghost-logical statement; and those, while poor substitutes for logical statements, are at least always intelligible, being usually just historical statements about scientists. But when logical expressions are embedded in volitional contexts, the result is simply unintelligible, at least to 'ordinary philosophers'. The word "unfalsifiable", for example, means, in Popper and all our authors, "consistent with every observation-statement". To "make a proposition unfalsifiable by *fiat*", then, is to make it, by fiat, consistent with every observation-statement. But how can anyone, whether the majority, or scientists, or even Popper, *make* it so? Or *make* the logical relation between propositions, in any other case, to be anything? Logical relations, surely, simply are not subject to the will.

Of course one can decide or choose what proposition a

given *sentence* will on a given occasion express. If I decide, as I can decide, that the next time I utter the sentence "The cat sat on the mat", the word "cat" will mean "bat", then that sentence may not on that occasion express the same proposition as it usually does. But it is certainly not this uninteresting and ever-present possibility that our authors have in mind. What they imply, and what they undoubtedly mean, is that the logical relation between propositions can properly be spoken of as made, decided, chosen, or the like.

If this were intelligible, it would be inexplicable how Popper ever allowed his original problem, the inconsistency of (1), (2) and (3), to arise. He 'solved' it, it now appears, by proposing a rule that would make E inconsistent with H, or by permitting us to regard E and H as inconsistent. But if he can do this kind of thing, why did he not do it in the first place? He could simply have permitted us to regard (1), (2) and (3) as consistent, or proposed a rule that would make them consistent. They aren't, of course, but then he would not be saying that they are: only exercising his will, permitting or proposing that they be so regarded. After all, he is permitting us to regard E and H as inconsistent, or proposing a rule that will make them inconsistent, though they aren't; and he is not saying they are, but only exercising his will and proposing a rule that will make them so. Why postpone the exercise of so sovereign a will?

But the major difficulty, as I said, is simply that of understanding our authors when they embed logical expressions in volitional contexts. The proposition H, again, is consistent with every observation-statement, including E. How then can any choice, decision, rule, or any operation of any will, even the divine will, make H and E inconsistent? This is, as James I said of the *Novum Organum,* like the peace of God, which passeth all understanding.

There is one important thing which is clear, however, and it is this: if a logical expression is embedded in a volitional context, then it is sabotaged. Thus, for example, it would not be possible for anyone to *make* P and Q inconsistent (say), or *to decide* that they are inconsistent, if they *are* inconsistent. With this much, I believe, even the most voluntarist of our authors would agree. But then it follows that if anyone *could* make P and Q inconsistent or decide that they are inconsistent, then P and Q would be *consistent,* not inconsistent. That is, a logical expression, once embedded in a volitional context, *cannot* retain its implication about logical relations, but must be sabotaged.

Every time, therefore, that our authors embed a logical expression in a volitional context, we have yet another instance in their writings of a logical expression being deprived of its implication about the logical relation between propositions.

T HE practice of sabotaging logical expressions by embedding them in contexts *about scientists,* and again Kuhn's 'tautological optimism' about science, correspond, in an easily understandable way, to a certain substantive *thesis* about logical expressions and scientists. Namely, that in the application of logical expressions, the highest authority, or at least some special authority, resides in science; that is, that scientists, whether in virtue of their knowledge or in virtue of their will, have some special authority on statements of logic. If like Kuhn you cannot tell the difference between (for example) the logical statement "P entails Q", and the ghost-logical statement "Any scientist would regard P as entailing Q", then you will think it out of the question that on matters of logic there could be any authority higher than science, or even any authority independent of science. And if you think this, then, since a statement of what the logical relation is between two propositions is equivalent to certain statements about how rationally conclusive certain inferences are, you will think that science has some special authority on questions of the rational conclusiveness of inferences.

It is interesting, therefore, to find that this thesis is actually affirmed by Kuhn. He writes: "To suppose that we possess criteria of rationality which are independent of our understanding of the essentials of the scientific process is to open the door to cloud-cuckoo land."[31]

Beliefs about the rational conclusiveness of inferences are, I take it, among 'criteria of rationality'. But if so, then Kuhn's thesis is not only false but the exact reverse of the truth. For we do possess, all of us, and by the million, criteria of rationality, and correct ones at that—that is, *true* beliefs about the rational conclusiveness of inferences—which are entirely independent of our understanding of science.

Everyone (or near enough) knows that "Socrates is mortal" is entailed by "All men are mortal and Socrates is a man". That the former does not entail the latter. That "Socrates is mortal" is less probable in relation to "Socrates is a man", than it is in relation to the conjunction of that proposition with "All men are mortal". That "All men are mortal" is more probable in relation to "Socrates is a man and Socrates is mortal", than it is in relation to the first conjunct of that proposition alone. And so on. Such logical knowledge may properly be called "natural", since everyone (near enough) possesses it. Obviously, too, everyone has enormous amounts of it. And it is clearly independent of the understanding of science. For such knowledge has been and is now possessed by a great many people who never so much as heard of science. To paraphrase Locke, God did not deal so sparingly with mankind as to make them barely two-legged, leaving it to scientists to teach them which inferences are rationally conclusive, or to what degree.

It may be said that natural logical knowledge, though it does not require acquaintance with *actual* science, nevertheless is knowledge of at least some of "the *essentials* of the scientific *process*": which were Kuhn's words. In a sense of course this is true: namely, in the sense that scientists do have natural logical knowledge and could not do their work if they did not. But in that sense natural logical knowledge is knowledge of at least some of the essentials of the legal process too, and of the haircutting process; for it is true of lawyers and barbers, too, that they have natural logical knowledge and could not do their work if they did not. Taken in this attenuated sense, then, Kuhn's thesis would be true. But then it no longer implies, what Kuhn appears to have meant by it, that on matters of how rationally conclusive inferences are, scientists have some authority which other people lack.

O UR authors, as I said in Chapter One, do not neutralise success-words all the time. Mixed strategy forbids it. Besides, they need to have these words on hand, with their success-grammar intact, as weapons to repel the *different* neutralisations which may be made of them by *other* people: people who are not licensed, as our authors are, to kill scientific success with words. It is all right for Popper to call scientific knowledge conjectural, but a mere undergraduate who in an essay called knowledge a poached egg, or said that knowledge entails falsity, would no doubt be given by our authors the sharp reminders he deserves about the meaning of some common English words. He might even be reminded (since this is clearly no time to be neutralising) that knowledge entails truth. Well, it is just the same with the sabotage of logical expressions. Our authors by no means do it all the time. Mixed strategy forbids that; and besides, they need to have logical expressions on hand, with their implications about logical relations intact, to use against other and unlicensed saboteurs of logic. They need them too, no doubt, for the humdrum task of correcting the mere errors that undergraduates make in their logic exercises.

Popper writes: "I had introduced in Chapter VIII of [*The Logic of Scientific Discovery*] a methodological rule permitting us to neglect 'extreme improbabilities'."[32] And he continues: "Certain sorts of coin behaviour [he means, for example, a million heads in a row] are incompatible with the coin's being fair (given our rule) and *that is that* ...".[33]

It is just as well Popper introduced this rule. Otherwise we might have gone on indefinitely just neglecting extreme improbabilities in our old bad way: that is, without his permission. But at least until Popper introduces a rule of higher type, permitting no one else to introduce other rules of the first type, other people may introduce other such rules. 'With one bound Jack was free' is a good game, but it has the drawback that any number can play.

Popper tells us that he has had students who thought at first that "All men are mortal and Socrates is mortal" entails "Socrates is a man"; but that he succeeded in getting them to acknowledge their error.[34] But such a student might not have been by any means so docile on this point, it seems to me, if he had read a little more widely in his teacher's writings.

He might have said: 'I have introduced, in my logic exercise, a methodological rule permitting us to deduce "Socrates is a man" from "All men are mortal and Socrates is mortal". Given our rule, this premise is incompatible with Socrates' not being a man; and *that is that*.'

Or suppose the subject were an *inductive* argument: say, "All the observed ravens have been black, so at least one of the unobserved ravens is black." Here Popper would urge upon the student his famous discovery, that such an argument is invalid. But why should the student tamely submit to this? If he wants to sabotage logical expressions, his teacher has given him ample precedent. And is he bound to sabotage only those logical expressions, and only in those places, that his teacher's example has previously authorised? Surely not. He might, then, choose a suitable epistemic context, and say for example, in the best ghost-logical style, *and* with plausibility: "Any ornithologist would regard that argument as valid." If he prefers sabotage by volitional embedding, he may simply *propose* a rule making this inductive argument valid; or perhaps his whole class, or the undergraduate body, will propose such a rule. Again the students would have Popper's precedent to justify them; and just as there is no special authority over logical relations attaching to Popper's will, so there is no special lack of authority attaching to undergraduate wills. More simply still, the students might make a point of writing, in all their essays, never that inductive arguments are invalid, but always that they are 'invalid'; and see what objection their teacher could consistently make to this practice.

A pupil so apt as this one would oblige Popper, or any of our other authors, to fall back on doing what ordinary philosophers do, or try to do, all the time. That is, to use only *un*-sabotaged logical expressions, and to talk plain English about the logical relations which exist between propositions, independently of scientists' or anyone else's 'regards', 'proposals', and bedding generally. There would then be no embedding of logical expressions in either epistemic or volitional contexts, no enclosing them in quotation-marks, in fact no sabotage of logical expressions at all; at least for a little while.

WHILE this lucid interval lasted, there would be a respite, too, from another maddening feature of the English of some of our authors: a feature which, though it is distinct from their sabotage of logical expressions, could not have come about but for that, and could not subsist if sabotage were dropped. I mean the practice, so common in Lakatos especially, of speaking in such a way as to weld logical and *causal* relations into one solid mass of confusion. For example, the application to scientific theories of expressions such as "is defeated", "is eliminated", "is removed", "is abandoned", as though these *causal* expressions were logical expressions like "is falsified", only perhaps stronger; and as though, consequently, the logical relation between propositions could have causal power.

It is very easy to see how this practice came in. P's falsifying Q cannot itself have any share of causal power; for example the power to cause abandonment of belief in Q. But of course P's being *regarded as* falsifying Q can have a share of causal power, just as the knowledge or belief that P falsifies Q can. And our authors, by their constant sabotage of logical expressions, have succeeded in blurring the distinction between P's falsifying Q and its being regarded as doing so. But the practice is one which is deplorably well-adapted to reinforce, at the same time as it expresses, the conflation of the history with the logic of science.

Some other expressions, and very important ones, which at least appear to confuse logical with causal relations, are those to which Popper and Feyerabend did most to give currency: "theory-dependence", "theory-ladenness", and their cognates. The thesis of the theory-ladenness of all observation-statements is by now of course widely accepted, and widely regarded as a major support of irrationalist philosophy of science. Whether it does support irrationalism, however, depends on what is logical, and what is causal, in this relation of ladenness or dependence. Theories and observation-statements are both propositions, and the relation of ladenness is evidently one which, at least in part, depends for its existence on the content of the propositions it relates. So far, then, ladenness seems to be a logical relation, and could even be a purely logical one. On the other hand it sometimes seems to be a purely causal relation. For proponents of the theory-dependence of all observation sometimes take it as sufficient to establish that the observation statement O depends on the theory T, that a scientist could not, causally speaking, have recognised the truth of O had he not at least entertained T. Yet this can hardly be seriously intended. For obviously it might also be true that a scientist could not, causally speaking, have recognised the truth of O had he not been in good health, or a member of the physicists' trade-union; while no one would take *that* as sufficient to establish that the observation-statement O is health-laden, or trade-union dependent. (On second thoughts, perhaps Kuhn *would*.)

There may then be an interpretation of the thesis of the theory-dependence of observation in which it is true and does not support irrationalism. My own suspicion is that there is not, and that on the contrary "Observation-statement O depends on theory T" is always just a ghost-logical statement in an indeterminate or foetal stage of development, and that the right regimen for it is abortion or exposure. Anyway, for the opponents of irrationalist philosophy of science there is nothing more urgently required than to focus critical attention on this quasi-logical/quasi-causal relation of dependence or ladenness, of which so much has lately been made.

APPENDIX TO CHAPTER TWO

Helps to Young Authors (II)
Sabotaging logical expressions

Q: This is a fair coin.
P: It has just been fairly tossed 1000 times and it came
     down "heads" 900 of those times.

How to rewrite the sentence: Q is consistent with the true
observation-statement P, but very improbable in relation to it.

Lakatos:	Q is 'falsified' by P.
Popper: (Either)	Scientists, by a methodological convention, regard Q as falsified by P.
(or)	I introduce a methodological rule permitting us to regard P as incompatible with Q.
Kuhn:	A scientist who did not, in view of the anomaly P, reject Q, would be regarded as violating one of the most basic values of his profession: consistency.
Feyerabend:	The theory Q, though 'refuted' by the anomaly P and a thousand others, may nevertheless be adhered to by a scientist for any length of time; and 'rationally' adhered to. For did not the most 'absurd' of theories, heliocentrism, stage a come-back after two thousand years? And is not Voodoo now emerging from a long period of unmerited neglect?

PART TWO

HOW IRRATIONALISM ABOUT SCIENCE BEGAN

CHAPTER THREE

THE HISTORICAL SOURCE
LOCATED

POPPER, Kuhn, Lakatos and Feyerabend have succeeded
in making irrationalist philosophy of science acceptable
to many readers who would reject it out of hand if it
were presented to them without equivocation and consistently.
It was thus that the question arose to which the first Part of
this book was addressed: namely, how did they achieve this?
My answer was, that they did so principally by means of the
two literary devices discussed in Part One. The question to
which the present Part of this book is addressed is: how was
irrationalist philosophy of science made acceptable *to these
authors themselves?*

Some part of the answer to this question no doubt lies in
those very misuses of language which have already been
discussed. For there is no reason to suppose that our authors'
characteristic treatment of logical expressions and success-words
has imposed on the writers any less than on their readers. But
obviously there must be some much more basic answer than
this to the historical question which I have just raised. How
did irrationalism about science come to recommend itself *at
all and in the first place,* to some leading philosophers from about
1920 onwards, as it did not, and could not have done, to their

counterparts a hundred or two hundred years earlier? It must be in principle possible to explain this phenomenon, just as it is possible in principle to explain any other large-scale movement in the history of thought.

It is not to be assumed, of course, that the origins of recent irrationalist philosophy of science are *purely* intellectual: that this philosophy came into being solely as a result of our authors accepting some thesis or other, and duly accepting its logical consequences. The common-sense assumption is in fact the other way. Any large-scale movement of thought is likely to be brought about, at least in part, by non-intellectual causes; and the present case is presumably no exception.

Nevertheless it will be taken for granted here that the origins of the movement of thought with which we are concerned are at any rate *principally* intellectual: that is, that the irrationalist conclusions of our authors' philosophy *are* embraced by them principally because they are logical consequences of some premises which these authors accept. Not to take this for granted would amount to intolerable condescension towards the authors in question, similar in kind to that by which Marxist writers 'explain' Darwin as though he were some simple mechanical toy.

The question is, then, what are the intellectual origins of recent irrationalism in the philosophy of science? Since we are looking for *intellectual* origins, the answer must consist in some *thesis* or other. And since we are looking for *origins,* the thesis must be one which functions in our authors' philosophy as a *premise,* and not as a consequence of other theses. Further still: what we seek to identify is that one among their premises which is *the key premise* of their irrationalism, in the sense that without it their philosophy of science would not have (that is, the other premises of it do not have) any irrationalist consequences at all.

Our question, then, is purely historical. The answer to it, however, is not of historical interest only. It would indeed be extremely interesting, as a matter of the history of thought, to know what is the key premise, in the sense just explained, of recent irrationalist philosophy of science. But the *philosophical* interest which indirectly attaches to our enquiry is greater still. What philosophers will want most to know, concerning the key premise of our authors' philosophy, is whether or not it is *true*. But in order for that to be known, it is an obviously indispensable preliminary, that it be known what this proposition is.

In this book only the preliminary and historical task, of identifying this proposition, is attempted; not the philosophical task of determining its truth-value. But if we can do even this much, then there will be some immediate and substantial benefit to philosophers. Controversies constantly take place between our authors (or their followers) and other philosophers who, while they share *some* of our authors' premises, disagree with their irrationalist conclusions. If our authors' key premise were once identified, then it would be known, to both sides in such controversies, where their disagreements begin. How valuable such information is, in enabling pointless debate to be avoided, and yet how hard to come by in philosophy, no philosopher will need to be told.

2

SINCE most of the quotations in Part One illustrated ways in which our authors' irrationalism is *disguised*, we should here satisfy ourselves that the phenomenon which we wish to explain really does exist: that is, that our authors' philosophy of science really is irrationalist. The best way to do this with reasonable brevity is to put before the reader (who is assumed to be familiar with their writings) a few concrete and pungent reminders of those writings: to cite some things our authors say about science which, while they are indisputably representative of their philosophy, are at the same time extremely and overtly irrationalist. This is what is done in the present section.

First, then: if there has been a great increase of knowledge in recent centuries, then *a fortiori* there sometimes are such things as positive good reasons to believe a scientific theory; but Popper says expressly, repeatedly, and emphatically, that there are not and cannot be such things. This thesis is so startlingly irrationalist that other philosophers, as Popper himself tells us, sometimes "cannot quite bring [themselves] to believe that this is my opinion". But it is: "There *are* no such things as good positive reasons"[1] to believe any scientific theory. "Positive reasons are neither necessary nor possible".[2]

These opinions will be admitted to be irrationalist enough: and they are too deliberately and emphatically expressed to be unrepresentative.

A scientific theory, Popper never tires of reminding his readers, is never certain in relation to, or in other words deducible from, those propositions which constitute (in most people's eyes) the reasons to believe it. Of course I do not cite *this* as an irrationalist thesis. It is only a *fallibilist* one: it asserts no more than the *logical possibility* of the conjunction of the evidence for any given scientific theory, with the negation of that theory. This thesis is so far from being one which is peculiar to the authors with whom we are concerned, that it is nowadays a commonplace with almost all philosophers of science. But Popper goes much further than this. It is a favourite thesis with him that a scientific theory is, not only never certain, but never *even probable,* in relation to the evidence for it.[3] More than that: a scientific theory, he constantly says, cannot even be more probable, in relation to the empirical evidence for it, than it is *a priori,* or in the absence of all empirical evidence.[4]

These two theses will be acknowledged to be irrationalist enough; and they are ones upon which Popper repeatedly insists. He goes much further still, however. The truth of any scientific theory or law-statement, he constantly says, is exactly as improbable, both *a priori* and in relation to any possible evidence, as the truth of a self-contradictory proposition[5]; or, to put the matter in plain English (as Popper does not), it is impossible.

Again: scientific knowledge is usually thought to have at least some connection with rational belief, but Popper writes: "Belief, of course, is never rational: it is rational to *suspend* belief".[6] One hardly knows what to wonder at more here, the thesis itself, or the arrogance of the author's "of course". His thesis, as will be evident, goes far beyond the philosophy of *science.* But it certainly does go as far as that, and will be admitted to express, in that domain, an irrationalism sufficiently uncompromising.

Again: Popper endorses the notorious sceptical thesis of Hume concerning inductive arguments, or arguments from the observed to the unobserved. This is the thesis that no proposition about the observed is a reason to believe any contingent proposition about the unobserved; or in other words, that the premise of an inductive argument is never a reason to believe its conclusion. Popper constantly and emphatically, and with detailed references to Hume, expresses his assent to this thesis. He writes, for example: "I agree with Hume's opinion that induction is invalid and in no sense justified."[7] And again: *"Are we rationally justified in reasoning from repeated instances of which we have experience to instances of which we have had no experience? Hume's unrelenting answer is: No, we are not justified... My own view is that Hume's answer to this problem is right..."*[8]. There are many other statements by Popper to exactly the same effect.[9]

Scepticism about induction is an irrationalist thesis itself, but its irrationalist character is enormously amplified if it is combined, as it is combined in Hume and in Popper, with the thesis of empiricism: that is, with the thesis that no propositions *other than* propositions about the observed can be a reason to believe a contingent proposition about the unobserved. For then it follows at once (since inductive scepticism says there can be no reason from experience) that there can be no reason *at all,* to believe any contingent propositions about the unobserved: which class of propositions includes, of course, all scientific theories. Hume, being an empiricist, did draw from his inductive scepticism this even more irrationalist conclusion: 'scepticism about the unobserved', as we may call it. And Popper, for the same reason, does the same.

Hume's inductive scepticism, while it is one irrationalist thesis among others in Popper's philosophy of science, is also more than that: it is one on which all the others logically depend. Whenever Popper undertakes, as he often does, to explain the grounds of his philosophy of science, and especially of whatever is most irrationalist in it, the reader is sure to meet with yet another of Popper's expositions, with detailed reference to Hume's writings and with unqualified endorsement of Hume's scepticism about induction.[10] If we take any other representative expression of Popper's irrationalism (for example those mentioned above in the second to the sixth paragraph of this section), and ask ourselves "Why does Popper believe this?", then part at least of the answer is always the same, and always obvious. It is because he shares Hume's scepticism about induction.

It would be easy to extend indefinitely a list of irrationalist theses which are representative of our authors; but there is no need to do so here. The examples given above suffice for the present purpose, which was only to satisfy ourselves that the philosophy of science here in question really is irrationalist. It is a sufficient condition for a philosophy of science to be irrationalist (as was said at the beginning of this book) if consistency with it requires reluctance to admit that there has been a great increase of knowledge in recent centuries. Popper's philosophy of science, it will be evident even from the few samples of it given above, fulfils this condition amply.

The examples of irrationalist theses given above were not only few in number, but were all drawn from Popper, none of them from any of our other three authors. But this too is perfectly proper, and in fact appropriate. Popper's philosophy of science is at any rate not more irrationalist than that of Feyerabend, Kuhn, or Lakatos, and at the same time, as a matter of well-known history, Popper's philosophy owes nothing to theirs, while Kuhn's philosophy owes much, and the philosophy of Lakatos or of Feyerabend owes nearly everything, to Popper.

3

OUR object, then, is to identify the key premise (in the sense explained earlier) of the reasoning by which our authors have been led to such irrationalist conclusions about science as have been cited in the preceding section.

There is no reason to expect this identification to be very easily made. It is always harder to identify a person's premises than to identify his conclusions. The reason is obvious. A reasoner's premises or starting-points are those propositions which he feels most entitled to take for granted. They are, therefore, the parts of his reasoning which are least likely to be explicit enough to enable other people to identify them easily. Indeed, it is sometimes difficult or even impossible for the reasoner *himself* to identify all his premises. For a proposition can be a premise of a person's reasoning without his ever having put it into words, and even without his being conscious of believing it at all.

It is nowhere of more importance than in philosophy to make clear what our reasoning is, and hence what our premises are; and most philosophers, accordingly, at least aim to achieve these things. But, whether from differences in temperament or in training, their actual achievements in this respect are very unequal, and many philosophers simply are not clear enough reasoners to enable their premises to be identified with any confidence. Again, it will be difficult to identify a philosopher's premises, however clear a reasoner he may be, in proportion as his philosophy is derivative from some one else's. If, for example, what one philosopher does is principally just to *illustrate* a position which he takes to have been placed beyond dispute by another philosopher, then it will hardly be possible to discover, from *his* writings, what the ultimate grounds are on which that position rests.

For these reasons, it would be idle to try to identify the key premise of recent irrationalist philosophy of science, from the writings of Lakatos, Feyerabend, or Kuhn. Lakatos is the only one of these three who is a clear enough reasoner to hold out any hope of such identification. But it is in fact impossible in all three, because of the extremely derivative character of their philosophy. In their writings, irrationalism about scientific theories functions, not as a conclusion at all, but as a premise, and as an inexplicit and scarcely-conscious one at that. Of what such irrationalism is a *consequence,* it is the least of their concerns to make clear. They are hardly to be looked to even for the *enunciation* of general irrationalist theses about science, such as Popper scatters so freely over his pages; still less, therefore, are they to be looked to for the arguments for them. In recent irrationalist philosophy of science, these authors are *fortunate heirs,* and like most persons of that kind, they are more concerned to enjoy their inheritance than to enquire into the grounds of it. Feyerabend and Kuhn made some slight additions to their irrationalist inheritance; Lakatos made some trifling abridgments of it, as though he were slightly uneasy about it; but what all of them chiefly did was simply to *illustrate* it, from chosen episodes in the history of science.

Popper on the other hand, writing as he was a generation before these authors, and for a less enlightened age, was obliged, as they never were, to work for his irrationalist theses: to *argue* for them. He it was in fact, and no one else, who made 'straight in the desert a highway' for these writers, so that irrationalism could thereafter be treated as a settled thing and a starting-point. it is to Popper, therefore, and to him alone, that we must look, in our attempt to identify the key premise of recent irrationalism. But since he is also a clearer reasoner than any of our other authors, we can do so with some prospect of success.

IN such theses as those of Popper which were mentioned in the preceding section, there is nothing new. What were there cited as representative expressions of new irrationalism, could equally be cited as representative expressions of old scepticism. That it is always rational to suspend belief, is a thesis of Pyrrho as well as of Popper: that from what has been experienced, nothing can be rationally inferred about what has not, is a thesis of Hume as well as of Popper; and so on. It *is* new, of course, to have such sceptical or irrationalist theses as these filling huge books called "The Growth of Scientific Knowledge"[11], "The Logic of Scientific Discovery", etc., etc. But then (as was said at the beginning of this book), when it is obvious that knowledge has increased, authors who wish to imply the opposite and yet retain plausibility must write in ways apt to mislead their readers. But in the substance, as distinct from the literary form, of Popper's philosophy, nothing is new.

In particular, Popper himself makes clear (as I have said), that the scepticism of Hume about inductive arguments is not only one of his own irrationalist theses, but part of the immediate grounds of all the others.

In this dependence on Hume, Popper is only an extreme case of a general condition. For the influence of Hume on 20th-century philosophy of science in general is in fact so great that it is scarcely possible to exaggerate it. He looms like a colossus over both of the main tendencies in philosophy of science in the present century: the logical positivist one, and the irrationalist one. His empiricism, his insistence on the fallibility of induction, and on the thesis which follows from those two, of the permanent possibility of the falsity of any scientific theory, are fundamental planks in the platform of both of these schools of thought. Where the two schools separate is that the irrationalists accept, while the logical positivists reject, Hume's further, *sceptical,* thesis about induction: that the premise of an inductive argument is no reason to believe its conclusion. This is why the logical positivists, in the 1940's and '50's, set about constructing what they called 'confirmation-theory', 'non-deductive logic', 'the theory of logical probability', or 'inductive logic': a branch of logic which, while being consistent with empiricism and inductive fallibilism, would allow scientific theories to be objects of rational belief without being *certain.* The irrationalists, on the other hand, being Humean sceptics and not merely fallibilists about induction, deny the possibility of any such theory; and Popper, accordingly, makes the chief landmark of 'inductive logic', Carnap's *Logical Foundations of Probability,* a principal target of his criticism.[12]

In the sharpest possible contrast to all this, the influence of Hume on philosophy of science in the 19th century was but slight. For this extraordinary reversal in the importance attached to Hume's philosophy of science, the historical reason is obvious enough, at least in broad terms. The crucial event was that one which for almost two hundred years had been felt to be impossible, but which nevertheless took place near the start of this century: the fall of the Newtonian empire in physics. This catastrophe, and the period of extreme turbulence in physics which it inaugurated, changed the entire climate of philosophy of science. Almost all philosophers of the 18th and 19th centuries, it was now clear, had enormously exaggerated the certainty and the extent of scientific knowledge. What was needed, evidently, was a far less optimistic philosophy of science, a rigorously *fallibilist* philosophy, which would ensure that such fearful *hubris* as had been incurred in connection with Newtonian physics should never be incurred again. Well, the very thing needed was lying at hand, though long neglected; and Hume, 150 years after his death, finally and fully came into his own.

Thus the revival of Hume's philosophy of science in this century was a movement of retreat from that confidence in science which was so high, and constantly rising, in the two preceding centuries, and which had proved to be misplaced precisely where it was highest. This retreat was general, all empiricist philosophers taking part in it. Popper and his followers are simply those with whom the retreat turned into a rout. *They* fell back *all* the way to Hume: not just to his fallibilism but to his *scepticism* about induction; and hence (since they were empiricists) to his scepticism in general about the unobserved.

Their only object was, and has remained, to ensure that no scientific theory should ever again become the object of over-confident belief; since only in that way can it be guaranteed that such a fall as overtook Newtonian pride will never be repeated. Now, it was the belief that a scientific theory can be *certain,* which had made that fall possible. So it must be re-affirmed, with Hume, that a scientific theory is *never deducible from* the observational evidence for it. On this negative logical relation Popper and his followers accordingly insist, and insist *ad nauseam,* even though no empiricist any longer dreams of denying it. They insist on it to the exclusion of every other logical relation which might exist between a scientific theory and the evidence for it, and they deny, with Hume, that propositions about the observed can ever be a positive reason to believe a scientific theory. They must do so: otherwise it might one day happen that a scientific theory should again be mistaken for a certainty. And that, for these philosophers, is what must at any cost be prevented.

This same consuming anxiety, it is worthwhile to point out, finds expression even in the very germ of Popper's philosophy: that is, in his opinions as to what *constitutes* a scientific theory, and what makes one such theory better than another. The very mark of a scientific theory, he thinks, is that it should be able to be *disproved* by experience[13]; and one scientific theory is better than another (other things being equal), he thinks, if it is more disprovable than the other.[14] No opinions could express more poignantly than these the depth of Popper's dread lest Newtonian *hubris* should ever have a sequel. For this is to say that the very mark of a scientific theory is that it be possible for us to *repel* any claims it might have on our belief, and that a theory is the better, the *more easily* the burden of belief which it threatens to impose on us can be put off. And nothing, evidently, could have suggested so strangely inverted a conception of science, except the most intense recollection of the traumatic consequences of having once fully believed a false theory.

Such is the genesis of Popper's philosophy of science. It is a story of one kind of reaction to the disappointment of extreme expectations: that kind of reaction, namely, of which the best epitome is given in Aesop's fable of the fox and the grapes. The parallel would be complete if the fox, having become convinced that neither he nor anyone else could ever succeed in tasting grapes, should nevertheless write many long books on the progress of viticulture.

We have made a beginning, then, in our attempt to identify the key premise of recent irrationalist philosophy of science. That premise is to be looked for, among our authors, in Popper and nowhere else. The irrationalism of Popper about scientific theories has turned out to be no other than the scepticism of Hume concerning contingent propositions about the unobserved. We know what are the immediate grounds, both in Hume and in Popper, of that irrationalism or scepticism: the conjunction of the theses of empiricism and inductive scepticism.

It is obvious, furthermore, which of these two immediate grounds is the *key* to the irrationalism of this consequence of their conjunction. It is the thesis of inductive scepticism. From the empiricist ground on its own no irrationalist consequence follows.

But all this is only a beginning, since what we have so far identified are only the immediate grounds of Popper's irrationalism concerning scientific theories. What we want to know, however, are the ultimate grounds of it. At least, we want to know that ultimate ground without which his philosophy of science would have no irrationalist implications.

The thesis of inductive scepticism cannot possibly be *itself* that ultimate ground or premise of Popper's irrationalism. It operates as a tacit premise, indeed, in the philosophy of Feyerabend, Kuhn, and Lakatos; but then, that is just the principal respect in which these philosophers are the careless beneficiaries of Popper's labours. At no earlier period than theirs in the entire history of philosophy could a respected philosopher (not to say a sane man) have started from the *assumption* that the observed can furnish no reason to believe anything about the unobserved. Certainly Popper, writing in an earlier and less enlightened age, had to have some *argument* for so startlingly irrationalist a thesis.

Our search for the key premise of our authors' irrationalism leads us, then, to the question: what are the premises of Popper's argument for scepticism about induction? How was inductive scepticism itself established?

Just as in general our other authors are derivative thinkers in relation to Popper, so Popper in turn, here at any rate, is a derivative thinker in relation to Hume. Indeed, on this all-important matter of the grounds of inductive scepticism, he is entirely so. Popper's argument for scepticism about induction is simply Hume's argument for it. He has neither fault to find with Hume's reasoning for this conclusion, nor anything to add to it. "I regard Hume's formulation and treatment of the logical problem of induction ... as a flawless gem."[15] What Hume gave us, Popper says, is "a gem of priceless value... : a simple, straightforward, logical refutation of any claim that induction could be a valid argument, or a justifiable way of reasoning."[16]

This being so, we know at any rate this much about the key premise of Popper's argument for inductive scepticism: that it is the key premise, whatever that is, of Hume's argument for the same conclusion. For these arguments are one and the same.

The reader of Popper is naturally led to expect, by such passages as have just been quoted, that he is about to be told what this perfect and simple argument of Hume's was. But the reader is disappointed in this expectation. Hume's conclusion is there stated and endorsed by Popper, but his argument for it is only praised, not stated. There are, however, other places in his writings where he does attempt to say what Hume's argument was.[17] These accounts differ widely in how much of the detail of Hume's argument they disclose. Some of them are mere hints of the argument, too brief or obscure to make any of its internal structure visible at all.[18] In other cases Popper's account does succeed in making some of the structure of Hume's argument clear.[19] For our purposes, however, what is required is an account of Hume's argument which enables us to identify its *premises,* and *all* of them. From this point of view all Popper's accounts of Hume's argument are extremely deficient. It would be only with the greatest difficulty, if at all, that anyone could learn from Popper what even *one* of Hume's premises was.

It should not surprise us that Popper has reproduced only very incompletely the argument which he praises so lavishly. On the contrary, this was to be expected. It is simply another instance of that obvious rule which was stated earlier: that the more derivative one thinker is in relation to another, that is, the more he regards that other as having placed a certain conclusion beyond dispute, the less likely he is to make clear what the original grounds were on which that conclusion rested.

The deficiencies of Popper's account of Hume's argument do not, however, impose any obstacle to our enquiry. They are simply an additional reason why the historical focus of that enquiry must now go back beyond Popper. We must simply identify that premise of *Hume's* argument for inductive scepticism, without which it would not have its irrationalist conclusion. The fact that Popper's accounts of that argument are very imperfect, does not matter at all. Had he given ever so good an account of it, still, since the argument in question, by Popper's own testimony, is Hume's, it is Hume's argument to which we ought to turn our attention.

The shift of the focus of our enquiry back to Hume, while it is in any case necessary, is also attended by marked advantages. For one thing, Hume is a clearer reasoner than any of our four modern authors. Secondly, and even more important, the circumstances in which Hume argued for inductive scepticism were much more conducive to explicitness of argument on this point than those in which Popper did. Popper did so in a period of catastrophic collapse of confidence in science (as well as of confidence in much else)[20], a period in which irrationalist theses, such as inductive scepticism, were greedily embraced by many of his readers almost faster than Popper could write them down. Hume, living in a less enlightened age, had no such assistance. On the contrary, he had to argue for scepticism about induction, not only from a standing start (as it were), but entirely *against* the prevailing current of opinion. The current of Newtonian confidence, in particular, was already then so strong as to be irresistible except by the hardiest of sceptics. Popper, therefore, even if his native talent for clear reasoning had been as great as Hume's, was bound to be, on this subject, the less explicit reasoner of the two.

THAT Hume's philosophy of science is the source of a great deal of subsequent irrationalism, has been, of course, widely recognised: for example, by Bertrand Russell.[21] Indeed, it is emphasised by Popper himself.[22] Popper does not admit, of course, that his own philosophy of science is irrationalist, but it is as obvious to him as it is to everyone else that Hume's is[23], and he has been admirably explicit (as we have seen) in acknowledging the debt he owes to Hume.

In this respect our other authors compare very unfavourably with Popper. Their debt to Hume's philosophy (which means in the end, as we have seen, their debt to his sceptical thesis about induction) is not less than Popper's; it is only less direct. Yet one would look in vain in their writings for any direct, indeed almost for any indirect, acknowledgment of this indebtedness. Indeed, one has only to recall the thesis to which they are indebted (namely, that the premise of an inductive argument is no reason to believe its conclusion), to see at once how utterly out of place it would have been for these authors even to mention it. Popper had made Humean scepticism about induction so much *de rigueur*, that even to affirm it had become extremely unfashionable; almost as much so, indeed, as to deny it. For the authors of *The Structure of Scientific Revolutions,* or of *The Methodology of Scientific Research Programmes,* to introduce this simple old thesis into *their* works, would have been felt as an intolerable piece of rusticity. The proprietor of a pornographic bookshop may be dimly conscious of a debt to the author of *Areopagitica,* but Milton is the last person he wants to see in his shop.

In later works, however, there are two small and indirect indications that these authors do after all recognise, in this homely thesis of Hume, the progenitor of their own irrationalism.

Lakatos's philosophy of science was no sooner published than it was outflanked on the left (so to speak) by the still more irrationalist philosophy of his friend Feyerabend. Thus by a manoeuvre not the less amusing for being familiar, Lakatos found himself placed, late in his life, in the unaccustomed role of *defender* of science, *against* neo–Popperian irrationalism. In this extremity (we are told by Feyerabend, who is here referring mainly to unpublished discussions between them), Lakatos was reduced to objecting that even irrationalist philosophers do not "walk out of the window of a 50-storey building instead of using the lift. Towards the end of his life this seemed to be his main objection to me."[24] Feyerabend admits he was baffled by this objection "for quite a while"; as anyone might have been, by an objection so extremely *recherché*. Finally, however, he found a reply which the irrationalist can make to it, and he gravely explains what it is. This reply is fully as original as the objection, and is in fact, though apparently all-unknowingly, pure Hume. It does not matter, Feyerabend tells us, what he or anyone else "does or does not do", or feel, about walking out of high windows; what matters is that neither he nor anyone else "can give *reasons* for his fear" of doing so.[25]

Kuhn provides a less picturesque but equally clear belated acknowledgment of the central part played in his philosophy of science by scepticism about induction. In an article first published in 1977, he tells us that, if he finds himself unable to avoid certain views of science which some people regard as irrationalist, "that is only another way of saying that I make no claim to have solved the problem of induction."[26]

The ordinary philosopher comes across these two passages with mingled relief, astonishment, and indignation. Relief, because what he had privately believed all along, he now finds indirectly admitted, and admitted by the emperors themselves: that they have no clothes at all, except such as are woven out of Hume's scepticism about induction. Astonishment and indignation, because previously and apart from these two passages, nothing in these authors had prepared him for such an admission, and everything had in fact pointed the other way. There is *not one word* in Kuhn's *The Structure of Scientific Revolutions* from which a reader could infer that Kuhn believes that a problem of induction exists; much less infer that he believes it to have something to do with his philosophy of science. As for the debate about the rationality of believing one can safely walk out of high windows: what is this 'pastoral-comical' scene, in which Lakatos plays Beattie to Feyerabend's Hume, but an admission that what is principally at stake between irrationalists and their critics is the sceptical thesis of Hume about the possibility of learning from experience? A thesis which was old when Sextus Empiricus wrote, and which requires for its discussion examples no more esoteric than Hume's own one about walking out of windows[27], or the one always associated with Pyrrho, of walking over cliffs[28]! "But until now", the indignant reader exclaims, "these authors had led me to believe that, before I could enter the lists against their philosophy of science, I would have to have read at least as much as they have written about Galileo and the telescope, about Lavoisier and oxygen, about the Bohr-Kramers-Slater theory, about the Lummer-Pringsheim experiments, etc., etc. What! Was all of this really quite inessential all along? Was it bestowed on me, then, principally *ad terrorem?*" Alas, poor reader, it was.

However belated or infrequent their own acknowledgments of it, then, the philosophy of these authors depends, no less critically than the philosophy of Popper does, on the scepticism of Hume about induction. This historical fact has some extremely curious corollaries. For example, that had it not been for the author of the most famous of all attacks on the credibility of miracles, the author of *Against Method* would not have believed a vulgar charlatan who claimed to become a raven from time to time.[29] But it is in any case a fact, and we must now turn to the argument of Hume on which this all-important thesis of irrationalist philosophy rests. For the key premise, whatever it is, of Hume's argument for inductive scepticism, is also the key to our whole enquiry.

CHAPTER FOUR

THE KEY PREMISE OF
IRRATIONALISM IDENTIFIED

As a preliminary to identifying Hume's key premise, some explanations are needed. One of them concerns the way the word "inductive" is used throughout this book. It is the more necessary to say something of this, because the word is one which Hume himself never used in print.

A paradigm of "inductive argument", as that phrase is used here, is for example the argument from "All the many flames observed in the past have been hot", to "Any flames observed to-morrow will be hot too". (This example is based on one of Hume's.) Another paradigm is, the argument from the same premise to "All flames whatever are hot"; another is, the argument from the same premise to "Any flames on Mars at this moment are hot".

Inductive arguments can be of many other forms than these, and of much more complex forms. But these few simple examples will suffice to indicate that "inductive argument" is used here in exactly the same way as it has been generally used by philosophers since Bacon, and in exactly the same way as (for the most part) philosophers use it still. Induction is argument, as the traditional philosophical phrase has it, "from the observed to the unobserved". In an inductive argument, the premises are simply reports of something which has been (or could have been) observed; the conclusion is a contingent proposition about what has not been (and perhaps could not be) observed. In addition, of course, what the conclusion of an inductive argument says about the unobserved is *like* what the premises say about the observed.

It will be evident from the preceding paragraph, and especially from its last sentence, that the established philosophical concept of inductive argument is not a very well-defined one. Nevertheless, and perhaps surprisingly, philosophers have found the concept well-defined enough for all their purposes. A sufficient proof of this is the fact that in all particular cases—that is, once the premises and conclusion of an argument have been specified—philosophers never have any difficulty in reaching agreement as to whether the argument is an inductive one or not.

There is one aspect of the established sense of "inductive" which nowadays needs to be emphasised, because a sense of the word which is opposite in this respect has grown up in the last forty years. This is that, applied to arguments, it is a purely *descriptive* epithet. To call an argument or a class of arguments "inductive", is not *to evaluate* it at all. In particular, it is no part of what is meant by calling an argument "inductive", that its conclusion does not follow from its premises. Inductive arguments are simply a certain class (though indeed, if empiricism is true, a peculiarly important class) of arguments, distinguished from other classes by the fact that their premises and conclusions are propositions which respectively satisfy certain purely descriptive conditions.

Whether all or only some of the members of this class, or none of them, are *reasonable* arguments; what degree of logical value, if any, such arguments have: this *evaluative* question can, of course, be asked about inductive arguments, as it can be asked about any others. But no answer to it, or any part of an answer, is implied in simply calling an argument "inductive".

Philosophers have, of course, differed deeply in their answers to this evaluative question about induction. But before they can either agree or disagree about the logical value of a certain class of arguments, they need first to have a non-evaluative name for arguments of that class; and "inductive arguments", or "induction", is just such a name.

Now the scepticism of Hume concerning induction is one answer to the evaluative question which has just been mentioned. It is an answer to the question, what reasonableness or logical value, if any, inductive arguments possess; and it is an answer of the most negative kind. The premise of an inductive argument, Hume says, is no reason to believe the conclusion of it; a proposition about the observed is never a reason, however slight, to believe a contingent proposition about the unobserved. Hume, as I have said, does not himself call any arguments "inductive"; but the texts leave no room at all for doubt that, concerning those arguments which we call so, he embraced the thesis just mentioned. And this is his famous scepticism about induction.

Hume's philosophy "of the understanding" includes, however, very many other 'scepticisms' beside this one. Two of these require mention here, because there is some danger of their being confused with his scepticism about *induction*, though in fact they are quite independent of it. It is the latter alone, of course, with which we are concerned.

First, Hume's scepticism about induction must not be confused with what he calls "scepticism with regard to the senses".[1] This is a scepticism as to whether the senses really give us any access at all to the external world, even to those parts of it closest to us in space and time. There is a certain amount of this kind of scepticism, too, in Hume's own philosophy; but it is quite different from his, or anyone's, scepticism about induction. The latter is a denial of the reasonableness, *assuming* that the deliverances of the senses are to be believed, of believing on their account any conclusion which goes *beyond* them. It is, in short, a scepticism about *arguments from* premises of a certain kind, not about whether premises of that kind are ever available to us to begin with.

Second, Hume's scepticism about induction must not be confused with what he called "scepticism with regard to reason"[2]. This is, indeed, a kind of scepticism which is about arguments; and there is some of it, too, in Hume's philosophy. But the grounds Hume gives for his scepticism about induction are entirely independent of those he gives for his 'scepticism with regard to reason'; and the latter conclusion is no more about inductive arguments than it is about any other special class of arguments. Hume's 'scepticism with regard to reason' is, in fact, a denial of the reasonableness of *any kind of argument whatever:* logical or illogical, valid or invalid, mathematical or theological, empirical or ethical, philosophical or scientific whatever!

Later philosophers have taken little notice of the part of Hume's *Treatise* in which he defended this undiscriminating and (it must be admitted) uninteresting kind of scepticism; and of the little notice they have taken of it, most has been unfavourable. Hume himself was apparently willing enough, on more mature reflection, that his 'scepticism with regard to reason' should be forgotten. For although he is an exceptionally repetitive writer, and published the substance of the *Treatise* Book I again in the *Abstract,* and yet again in the first *Enquiry,* he never anywhere once mentioned *this* kind of scepticism again. By contrast, his argument for scepticism about *inductive* arguments is nothing less than the central thing in the *Treatise* Book I, the *Abstract,* and the first *Enquiry*. And how unwilling later philosophers, or at least twentieth-century ones, have been to forget *this* part of Hume's philosophy, we have already seen.

Hume argued for scepticism about induction (as has just been indicated) in three different books. They are *A Treatise of Human Nature*[3]; *An Abstract [of A Treatise of Human Nature]*[4] ; and *An Enquiry concerning Human Understanding*[5]. In each of these books Hume gives, for his sceptical conclusion about induction, only one argument. The argument is, however, the same each time. The three different versions of it differ only in conciseness, and in the degree to which the argument is mixed up with extraneous matter. But in these respects the three versions differ widely. In the *Treatise*, the relevant parts are Book I Part III Sections II-XIV. These Sections, which occupy almost a hundred pages (in the standard edition referred to in the bibliography), contain both *several* versions of the argument for inductive scepticism, and a great deal of other matter as well. The most concise version of the argument, and overall the best, is that given in the *Abstract;* where it occupies pp. 11–16 (of the standard edition referred to in the bibliography). In the *Enquiry* the argument is to be found, in a less concise form than that of the *Abstract,* but in a far more concise one than that of the *Treatise,* in Sections IV and V.

The account of this argument which is given below has grown out of an account which I gave in *Probability and Hume's Inductive Scepticism*[6]. There is only one respect of any importance in which these two accounts of the arguments actually conflict. To this I draw attention below, when I reach the relevant part of Hume's argument. But most of the many differences in detail between the two accounts are simply by way of addition. That is, much of the detail of Hume's argument, which was either left entirely unnoticed or at best suggested by my earlier account, is made explicit here. At the same time, the present account of the argument is intended to be, and I believe is, quite self-contained. In other words, while the reader, in order to judge whether the account given here of Hume's argument is correct and complete, will need familiarity with the parts of Hume's philosophy which were referred to in the preceding paragraph, he will not need anything else.

Hume's argument for inductive scepticism is itself, however, not quite self-contained. His thesis, that the premise of an inductive argument is no reason to believe the conclusion, is not quite the end of the argument in which it occurs. The argument for this thesis is only a part, though it is indeed by far the greater part, of Hume's argument for a sceptical conclusion which is far more general still. This is the Humean thesis which in the preceding chapter I called "scepticism about the unobserved". It says, there is no reason *whatever* (as distinct from merely, "no reason *from experience*") to believe any contingent proposition about the unobserved. It will be worthwhile to extend our account of Hume's argument so as to take in this, its very last, step; even though to do so involves some slight repetition of something which was said in Section 3 of Chapter Three above.

2

THE best place to begin is at the end of Hume's argument. The conclusion of the whole is a general sceptical thesis about whatever has not been observed: that there is no reason (from any source) to believe any contingent proposition about the unobserved. (Call this proposition *A*.) Here it is in some of Hume's own words: "we have no reason to draw any inference concerning any object beyond those of which we have had experience ..."[7].

There is no difficulty (as has already been indicated) in determining what Hume's immediate grounds are for this conclusion. For there are two propositions which he constantly asserts and clearly intends to be taken together, and which, when they *are* taken together, immediately and obviously entail *A*.

One of these grounds is empiricism: the thesis that the only reason to believe a contingent proposition about the unobserved is a proposition about what has been observed. (Call this *B*.) In some of Hume's own words: "... All the laws of nature, and all the operations of bodies without exception, are known only by experience...".[8]

The other immediate ground of *A* is Hume's inductive scepticism: the thesis that *even* propositions about the observed are not a reason to believe any contingent proposition about the unobserved. (Call this *C*.) In some of Hume's own words: " ... we have no argument to convince us, that objects, which have, in our experience, been frequently conjoined, will likewise, in other instances, be conjoined in the same manner...".[9]

The structure of this last step of Hume's argument is as easily identified as the elements of it. It was as represented in the following diagram.

$$\left. \begin{array}{c} C \\ \\ B \end{array} \right\} \rightarrow A$$

Our object being only to identify Hume's argument, not to evaluate it, the arrows in my 'structure-diagrams' are to be understood in a descriptive sense only, not in any evaluative one. Thus "X →Y", for example, would mean here, not that an argument from X to Y is valid, or that X is a reason to believe Y, or anything of that sort. It would mean that Hume in fact *gave* X *as* a reason to believe Y, and it would mean nothing else.

At the same time, it is quite obvious that Hume *intended*

his argument to be a valid one, and thought that it was. I too believe that the argument he intended is in fact valid. Of course Hume sometimes left unexpressed certain premises which are necessary to make his arguments valid, as every arguer must do if he is not to be tedious. But when he does, it is almost always easy, for a reader familiar with his philosophy "of the understanding", to supply the additional premise which Hume intended, and which is needed to make his argument valid. There is in fact only a single step in the entire argument of which this is not true. I intend to proceed, therefore, by assuming at each step that the argument is valid, and attributing to Hume the additional premise necessary and sufficient to make it so. Hence an arrow in my structure-diagrams, although it will not *signify* a valid step, will always represent a step which was (I believe) made by Hume, and which is (I believe) valid as well.

It will be obvious that Hume's scepticism about the unobserved (*A*) does follow from the conjunction of empiricism (*B*) with inductive scepticism (*C*). It will be equally obvious that *A* does not follow from *B* alone; even though many philosophers have thought, to the contrary, that Hume's scepticism about the unobserved is an inevitable consequence simply of his empiricism. It is not so obvious, but it is true and of some importance, that *A* does not follow from inductive scepticism (*C*) alone, either.

If you want to reach a certain place, it is no fatal news to be informed that the route *via* X will not get you there. This will be fatal news if and only if it is conjoined with the information that no route other than the one *via* X will get you there either. Just so, if you want to reach knowledge or reasonable belief about the unobserved, it is no fatal news to be told that the inductive route (the route *via* the observed), will not get you there. Yet that is all that inductive scepticism

C says. This will be fatal news if and only if it is conjoined with the information that no route other than the inductive one will get you there. Just that, however, is what empiricism B asserts. Hence scepticism about induction will not commit you to scepticism about the unobserved, unless you also subscribe to empiricism. Someone who held that there are non-inductive routes to knowledge or reasonable belief about the unobserved—routes *via* pure reason, say, or revelation— could with perfect consistency admit C and yet deny A: that is, be a sceptic about induction without being at all sceptical about the unobserved.

Neither B nor C was a *premise* of Hume's argument. Inductive scepticism C is, of course, so irrationalist a thesis that it could hardly be a *starting-point* of any argument advanced by a sane person (at any rate before about 1950); certainly *Hume* had to argue for it. But neither was empiricism B a starting-point of Hume's argument. For it, too, he argues. Hume's argument for B was sometimes perfunctory, it is true, as well as being usually short and elliptical. The historical reason for this is obvious: empiricism was a *commonplace* with Hume and with his readers. Hence B, quite unlike inductive scepticism C, was something which required little defence. Still, Hume does have an argument for empiricism. What was it?

Hume's main ground for empiricism, in the sense that it is the ground which he usually gives as though it were a sufficient one, is this: that propositions which are *necessarily true* are not a reason to believe any contingent proposition. (Call this D.) Unlike necessary truths, "matters of fact are not ascertained", Hume says, "by the mere operation of thought"[10], by "demonstrative arguments" or "abstract reasonings *a priori*". (When Hume speaks of "demonstrative arguments", he does not mean, as we might mean, just *valid* arguments; he means, valid arguments from *necessarily true* premises.[11])

It may appear from this that Hume begged the question in favour of empiricism. For the ground *D* just mentioned may seem scarcely distinguishable from the empiricism *B* for which it is supposed to be a ground. Well, it would not have been surprising, nor would it have mattered much at the time, if Hume's argument *had* been question-begging here, the reason being the historical circumstance mentioned a moment ago: that with Hume's contemporary readers empiricism was virtually a *datum* anyway. And since, as it happens, empiricism is virtually a *datum* with most of Hume's readers now, too, it would not matter much now, either, if his argument here had been circular. In fact, however, it was not.

Empiricism *B* says that if there is any reason to believe a contingent proposition about the unobserved, it is a proposition about the observed. Hume gives, as though it were sufficient to establish this, the ground *D,* that necessary truths are no reason to believe a contingent proposition about the unobserved (or any other contingent proposition). Now this is just like some one saying that the murder, if it was murder, was committed by the gardener, and giving, as though it were sufficient to establish this, the ground that at any rate the butler did not do it. Such a person is clearly assuming that the murderer, if there is one, is either the gardener or the butler. Equally clearly, Hume is assuming that if anything is a reason to believe a contingent proposition about the unobserved, it is either a necessary truth or a proposition about the observed.

This assumption, or rather the even stronger one, that any reason to believe *any* proposition is either a necessary truth or a proposition about the observed, is one which, once it is stated, will be acknowledged by every student of Hume to have been absolutely central to his thought. No account of his philosophy of the understanding can possibly be adequate if it does not make this assumption explicit and prominent. Without it, for example, it is quite impossible to explain Hume's special affinity with the empiricists of the present century: an affinity which (as was implied in Chapter Three above), is no less obvious than it is deep. And the deficiencies of my own earlier-published account of the present argument, I may observe, stem almost entirely from my having failed to make explicit the part played in the argument by this assumption.

The assumption has two parts, and it is helpful to separate them. Consider the class, at first sight the oddly disjunctive class, of propositions which are either necessary truths or propositions about the observed. What is common and peculiar to the members of this class? Or rather, what did *Hume* think is common and peculiar to them, and what gives them the special status that they enjoy in his philosophy? These questions are not hard to answer. Hume thinks of necessary truths and propositions about the observed as being propositions, and the only propositions, which can be known or reasonably believed, without having to be inferred from *other* propositions known or reasonably believed: as being propositions, and the only propositions, which are (as we may say) *directly accessible* to knowledge or reasonable belief. This is one half of his assumption. (Call it *E*.)

The other is a very natural assumption, about one proposition's being a reason to believe another, which is almost inevitably expressed (and is expressed by Hume) by means of a comparison with a chain or a ladder. P's being a reason to believe Q is like a ladder which reaches, whether by few rungs or many, from P to Q; and Hume's assumption is the exceedingly plausible one that such a ladder, no matter how safe and climbable it may be, will be no help at all to us for reaching Q, if we cannot reach P. In order, then, for P to be a reason, however remote or indirect, to believe Q, P must be *directly* accessible to knowledge or reasonable belief. Otherwise, as Hume says, all "our reasonings would be merely hypothetical; and however the particular links might be connected with one another, the whole chain of inferences would have nothing to support it…".[12] (The same condition is necessary, evidently, in order for P to be a member of a conjunction, P-and-R, which is a reason to believe Q.) (Call this second part of the assumption *F*.)

It is easy now to understand why Hume regularly proceeds as though

D: No necessary truth is a reason to believe any contingent proposition,

is sufficient to establish

B: Any reason to believe a contingent proposition about the unobserved is a proposition about the observed.

It is because he was assuming both

F: If P is a reason or part of a reason to believe Q then P is directly accessible to knowledge or reasonable belief,

and

E: A proposition is directly accessible to knowledge or reasonable belief if and only if it is either a necessary truth or a proposition about the observed.

The structure, then, of Hume's argument for empiricism, was:

$$\left. \begin{array}{c} D \\ E \\ F \end{array} \right\} \rightarrow B$$

I have implied that Hume's argument for empiricism *B* comes into his argument for scepticism about the unobserved *A*, only near the very end. So it does, logically speaking, since what it supplies is one of the immediate grounds of the ultimate conclusion. In the actual order of Hume's presentation, however, the opposite is true. He always completes the argument for empiricism B *first,* before he even begins the argument for inductive scepticism C. Moreover, when he does complete the latter, its conclusion is always a deliberate *echo* of a premise of the earlier argument for empiricism. C, the thesis that even after experience we have no reason to believe anything about the unobserved, echoes D, the thesis that we have no such reason *before* experience, or *a priori.* And Hume had good literary and historical motives for adopting this order of presentation of the parts of his argument, and in particular for adopting this echo-device.

Everyone dislikes a sudden loud noise, but it is worse still if you are half-asleep at the time. Now D, the thesis that we have no reason prior to experience to believe anything about the unobserved, is a proposition which I have elsewhere called "Bacon's bell"[13], in reference to Bacon's famous boast: that he had "rung the bell that called the wits together", by insisting, that all contingent propositions be subjected to the test of experience and to no other. But of course by the time Hume wrote, this empiricist maxim D, once so revolutionary, had become almost as much a part of the British constitution as a church by law established, and almost as soporific. So Hume, by sounding Bacon's bell *early* in his argument, as he always does, artfully creates in his readers a sense of security. Its familiar note assures them that this author is a decent British empiricist, a Bacon-and-Newton man like the rest of us: *he* will not disturb our Royal Society slumbers. How much the more appalling, then, when at the end of his argument he sounds what I have called "Hume's bell", with its ghastly parody of this familiar note: the thesis of inductive scepticism C, that we have no reason for any beliefs about the unobserved, *after* experience either!

To the main part of Hume's argument, his argument for

this staggering conclusion, we now turn. Here most of his premises are easily identifiable, and it is best to go straight to them.

One premise is, that the conclusion of an inductive argument does not follow from its premise, except in the presence of an additional premise, or assumption, that the unobserved is *like* the observed. In some of Hume's own words: "All inferences from experience suppose, as their foundation, that the future will resemble the past".[14] Again: "All probable arguments are built on the supposition that there is ... conformity between the future and the past ..."[15] Yet again: "...probability is founded on the presumption of a resemblance, between those objects, of which we have had experience, and those, of which we have had none..."[16].

Let us be sure we understand just what Hume is saying in these passages.

"Inferences from experience", "probable arguments", and "probability", are simply some of the many names which Hume uses for what we call inductive arguments: those arguments from the observed to the unobserved, of which the argument from "All the many flames observed in the past have been hot", to "Any flames observed tomorrow will be hot", may serve as a paradigm. And what Hume is pointing out is simply that this argument, for example, is invalid as it stands, (the conclusion does not follow from the premise), and that in order to turn it into a valid argument, you would need to add to it a premise which asserts at least that tomorrow's flames resemble the past observed ones.

Let us call a proposition which asserts that there is a resemblance between the observed and the unobserved, a "Resemblance Thesis". Then this first premise of Hume's argument for inductive scepticism is

G: Any inductive argument is invalid, and the weakest

addition to its premises sufficient to turn it into a valid argument is a Resemblance Thesis.

Hume's next premise is also easily identified. It is a proposition about the nature of Resemblance Theses; namely

H: A Resemblance Thesis is a contingent proposition about the unobserved.

In some of Hume's own words: "... that there is this conformity between the future and the past, ... is a *matter of fact*..."[17]

Now Hume concludes, from this characterisation of Resemblance Theses, something about the nature of possible evidence for them; namely

I: A Resemblance Thesis is not deducible from necessary truths.

In some of his own words: a Resemblance Thesis "can never be proved ... by any demonstrative argument or abstract reasoning *a priori*".[18]

It will be obvious that *I* does not follow from *H* alone, but equally obvious what Hume's tacit premise was here. It is the maxim, which he is never tired of repeating, that "there can be no demonstrative arguments for a matter of fact and existence". That is, it was

J: No contingent proposition is deducible from necessary truths.

This part of Hume's argument for *C* had, then, the structure:

$$\left.\begin{array}{c} H \\ \\ J \end{array}\right\} \rightarrow I$$

Hume next considers the possibility of *observational* proof of a Resemblance Thesis. But the result, he finds, is as negative as in the case of *a priori* proof. That is, he concludes, in an obvious parallel to *J*,

K: A resemblance Thesis is not deducible from propositions about the observed.

On what grounds? Well, recall *H*: that a Resemblance Thesis is a contingent proposition about the unobserved. Any argument to a Resemblance Thesis from *the observed* will thus be an *inductive* argument, and, in view of *G*, therefore, it will be invalid unless to the observational premises is added a Resemblance Thesis. That, however, is the very proposition which we are trying to prove! Any argument from experience for a Resemblance Thesis, therefore, will be invalid unless it is circular. Or in some of Hume's words, "To endeavour ... the proof of [a Resemblance Thesis] by probable arguments, or arguments regarding existence, must evidently be going in a circle, and taking that for granted, which is the very point in question".[19] Hume's grounds for *K*, then, are *H* and *G*. For they together entail

L: A Resemblance Thesis is deducible from propositions about the observed, only when to the latter is conjoined a Resemblance Thesis;

and *L* in turn (remembering that we already have *H* as a premise), is Hume's warrant for concluding that *K*.

Here, then, the structure of the argument was:

$$\left.\begin{array}{c} H \\ \\ G \end{array}\right\} \rightarrow L \rightarrow K$$

So far, then, Hume's argument for *C* has been as represented in the diagram:

$$\left.\begin{matrix} H \\ J \end{matrix}\right\} \to I$$

$$\left.\begin{matrix} H \\ G \end{matrix}\right\} \to L \to K$$

The only premises have been G, H, and J.

But now, what is represented above is, as far as I can discover, the whole of Hume's explicit argument for inductive scepticism C.

This assertion may be found surprising. Proof of it would certainly be desirable. Unfortunately, however, it is impossible. An old logical saw says that one cannot prove a negative, and certainly one cannot prove an exegetical negative such as this. I am obliged, therefore, to rely entirely here on what elsewhere Hume called "the method of challenge", and to invite anyone who thinks there is something explicit in Hume's argument for C, which is omitted in the above account of it, to point it out: either another premise, or a result drawn by Hume from premises already represented here.

If, as I believe, this cannot be done, then it must be admitted that Hume's argument, while it is admirably explicit as far as it goes, stopped a good deal short of the conclusion C which it was intended to prove. For C says that the premise of an inductive argument is not a reason to believe its conclusion; yet so far we have not got anything like that. In the premises G, H, and J (the only ones so far), there is nothing whatever, for example, about what is required for one proposition to be a reason to believe another.

Well, what *have* we got? Or rather, since it is inductive

arguments and no others which are the subject of C, we should ask how much, at the best, has so far been established about inductive arguments? The answer is plain. At the best (that is, assuming Hume's premises true as well as his steps valid), the most that follows from Hume's premises, about inductive arguments, is the conjunction of G, I, and K.

And all that that conjunction says is this: that any inductive argument is invalid, and that the weakest additional premise sufficient to turn it into a valid argument is a proposition which is not deducible either from necessary truths or from propositions about the observed.

We need to make this result less unwieldy. First, let us call any additional premise which is sufficient to turn a given invalid argument into a valid one, "a validator" of it; and let us call the weakest of all the validators of a given argument, *the* validator of it. Second, let us make use of the fact that, necessarily, a proposition is *deducible from* a necessary truth or a proposition about the observed, if and only if it is *itself* a necessary truth or a proposition about the observed. Because of this, instead of saying that the validator of an inductive argument is not deducible from necessary truths or observation-statements, we can say, without losing logical equivalence, that that validator is itself neither a necessary truth nor an observation statement. With the aid of these two abbreviations, what the conjunction of G, I and K says about inductive arguments can be expressed as

M: Any inductive argument is invalid, and the validator of it is neither a necessary truth nor a proposition about the observed.

Obviously, M does not entail inductive scepticism C. Indeed, since it is only an abbreviated logical equivalent of what G, I, and K say about inductive arguments, M cannot bring us any closer to C than the conjunction of G, I, and K did; which is, as I said, not very close. Many philosophers nowadays would go much further than this, and say that while *C*, though false, is at least important, *M*, though true, is unimportant, because its truth is obvious.

I wish that some of these philosophers would tell us an important and true result which *does* follow from the premises of Hume's argument: this famous argument which all of us (and not only those who accept its sceptical conclusion *C*) admire so much. In fact, however, the result *M* is not only true and original, but is of profound importance. Indeed it is Hume's central insight concerning induction, and is what separates his philosophy of induction, and the best of ours, from the slipshod philosophy of Bacon before and of Mill after him, and of most empiricists even now. Not only is *M* important in itself. In conjunction with some of the premises of Hume's empiricism (and ours), it entails, as will be shown later, a further result which is still more important, and one which most empiricists even now are far from having fully absorbed.

These are large claims to be made for the not-very-pregnant-looking result *M*; but I think I can establish them. To do so, however, it is necessary to step back from Hume's argument for a while.

Aᴌᴌ philosophers and all logicians are interested in evaluating arguments. The evaluation of arguments is a complex matter, requiring many different distinctions to be made. For example, in some arguments the premises *cannot* be a reason to believe the conclusion, while in other arguments they are, and hence can be. Then, where the premises are a reason to believe the conclusion, there is the distinction between arguments in which the premises are an *absolutely conclusive* reason to believe the conclusion, and those in which they are not; that is, between valid and invalid arguments from P to Q, where P is a reason to believe Q. Then there is the distinction, entirely independent of the two just mentioned, between arguments in which the premises are all true, and those in which they are not. And so on. In short, two arguments can differ in value along a number of different, and even independent, dimensions.

The ordinary or 'deductive' logician, however, is interested, *ex officio* at least, in only one dimension of the value of arguments: namely in the distinction between validity and invalidity. Most philosophers, on the other hand, regard the distinction between valid and invalid arguments as a silly thing to have as an *exclusive* object of interest. They are right. For one distinction which the evaluation of arguments requires to be made is, as I have said, that between arguments of which the premises are or could be a reason to believe the conclusion, and arguments in which they cannot; while that distinction is largely, if not entirely, independent of the distinction between the valid and the invalid. At any rate, it is certainly not enough to make P a reason to believe Q, that the argument from P to Q be valid.

If it *were* enough then no one, however irrational, need ever lack a reason, and even an absolutely conclusive reason, to believe any and every proposition whatever. For you can always turn an invalid argument into a valid one, merely by making a suitable addition to the premises. Let your argument from P to Q be invalid; let it even be as atrocious as a piece of reasoning can be; still, you can always turn it into a valid argument, by the trifling expedient of adding the premise that P is false or Q true, or some other premise which entails that one. Nothing could be easier. And if the conclusions following from the premises were enough to make those premises a reason to believe it, then nothing could be more important than this stratagem, since it would enable us all to ensure that whatever we believe, we believe reasonably.

In fact, of course, as is obvious, nothing could be more trivial. That the premises of an argument entail the conclusion is *not* enough to make them a reason to believe it. And if the premises of an argument are to succeed in being a reason to believe the conclusion, not every validator R of the argument from P to Q is available to every arguer as an additional premise. Such an R, to be available to an arguer as an additional premise, must at least be such that it *can* be part of a reason to believe Q.

Very often, of course, such a validator is available to an arguer. My companion may disagree with my identification of a bird which we are both looking at, and argue "The bird on that post is no raven, since all ravens are black"; omitting, just in order not to be tedious, the premise, which we have both just learnt from experience, that the bird on that post is not black. This proposition, which is of course a validator, in fact, *the* validator, of his argument, is available to him as an additional premise, presumably. At any rate it certainly satisfies a necessary condition of such availability: that of being a proposition which *can* be part of a reason to believe his conclusion.

But it is not so in every case, that is, for every invalid argument; and there are some validators which are *never* in any case available to arguers as an additional premise. If I am to succeed in giving a reason to believe a contingent proposition Q, but my argument to Q from P is invalid, I cannot add a premise R which is, for example, self-contradictory. A self-contradictory additional premise is indeed a validator of every invalid argument. But such a validator is not available to any arguer to Q, because a self-contradiction cannot be part of a reason to believe a contingent proposition. Again, if I aim to give a reason to believe Q, but my argument to Q from P is invalid, I may not add as a premise the very proposition Q which I am trying to give a reason to believe. The conclusion of any invalid argument is indeed a validator of it, but Q is not available to me or any arguer to Q as an additional premise, because Q cannot be part of a reason to believe Q.

Given an argument from P to Q which is invalid, then, a validator of it, R, may be available as an additional premise to an arguer whose object is to give a reason to believe Q. For R may be, in addition to whatever else is required for availability, a proposition which can be part of a reason to believe Q. In such a case, for example the argument about the bird on the post, the invalidity of the original argument is an unimportant defect of it, because a cure for the defect is available to the arguer. But not every validator R of a given invalid argument is available as an additional premise to every arguer whose object is to give a reason to believe Q. For a validator R may be a proposition which cannot be part of a reason to believe Q; or it may be a proposition which is unavailable on some other ground.

Hence for philosophers, who must distinguish, as deductive-logicians need not, between arguments in which the premises are or at least could be a reason to believe the conclusion, and arguments in which they cannot, an important general question arises. In *just what* cases is a cure for the invalidity of our arguments available to us, consistently with our premises remaining a reason to believe our conclusion? What propositions are, and what are not, available validators of our invalid arguments?

This question, in this general form, is not one which Hume ever explicitly considered. Still, we know well enough what his answer to it was, even in its general form. His answer to it is given by his premises *E* and *F* above. But Hume did of course consider, and most explicitly, the special case of this general question in which the arguments from P to Q are *inductive* ones. That is, he did consider the question whether, when we argue from the observed P to the unobserved Q, any validator R is available to us. Indeed Hume never considered any question, concerning the evaluation of induction, *except* this one. His answer to it is, of course, that *no* validator is available for inductive arguments. His argument for part of this answer is that which I have set out above.

What Hume did was to consider two classes of candidates for the position of available validators of induction. The first class consists of necessary truths. These were obvious candidates for consideration. Propositions which cannot be false are, presumably, *always* available as additional premises, to any arguer. At any rate they certainly satisfy the necessary condition of availability, that they *can* be part of a reason to believe the conclusion of the arguments now under discussion. When the argument from P to Q is inductive, a necessary truth R can be part of a reason, P-and-R, to believe Q.

Alas, where the argument from P to Q is inductive, a necessarily-true additional premise R, although available, will never satisfy the other requirement of the position we are seeking to fill; it will not be a validator. The conclusion of any inductive argument is a contingent proposition. Where R is a necessary truth, the conjunction P-and-R is logically equivalent just to P itself. And two arguments with the same contingent conclusion, and logically equivalent premises, cannot differ in value along any dimension (except perhaps an economic or an aesthetic one). At any rate they cannot differ in that one of them is valid and the other invalid. So where R is a necessary truth, an argument from P-and-R to contingent Q would be valid only if the argument to Q from P alone were valid to begin with; which, in the case of inductive arguments (as Hume's premise G says), it is not. Trying to turn inductive arguments into valid ones by adding necessarily true premises, is like trying to increase a boat's displacement by taking on weightless ballast.

Hume then considers a second class of candidates for the position of available validators of induction: propositions about the observed. These, too, were natural candidates for consideration. When we argue from the observed P to the unobserved Q, *another* proposition R about the observed is, presumably, very often available to us. Certainly such an R *can* be part of a reason, P-and-R, to believe a conclusion Q about the unobserved.

But alas, these candidates too are unequal to the task of turning inductive arguments into valid ones. By adding a proposition R about the observed, to the original premise P about the observed, the best you can get, that is at the same time a proposition which can be a reason to believe Q, is just another proposition about the observed; a stronger one, indeed, than that with which you began, but still a proposition about the observed. But the conclusion Q is still a proposition about the unobserved. So, even with the premise P-and-R, our argument to Q is an inductive one still. And all inductive arguments (as Hume's premise G says) are invalid. As far as turning inductive arguments into valid ones goes, therefore, propositions about the observed behave, as additional premises, *in exactly the same way as necessary truths.* At least *this* much is true, then, in the famous sceptical passage in which Hume writes: "Now where is that reasoning, which, from one instance, draws a conclusion, so different from that which it infers from a hundred instances, that are nowise different from that single one? ... I cannot find, I cannot imagine any such reasoning."[20]

No proposition, then, which is either a necessary truth or a proposition about the observed, is sufficient as an additional premise to turn an inductive argument into a valid one. *A fortiori* no necessary truth or proposition about the observed is the *weakest* of all the validators (that is, is *the* validator) of any inductive argument. That is, Hume's result M is true.

(It is, I hope, unnecessary to say that the argument just given for M was simply a modernised version of the argument, set out above, which Hume himself gave for it; and little enough modernised at that.)

I need not contend here for the originality of this result M. It is, in fact, as original as anything in philosophy ever is. What I do need to contend for is its importance. For, as I indicated earlier, many philosophers nowadays suppose that the truth of M is obvious, and even that it always was so, at least to philosophers. Some go as far as to suggest that M is an analytic truth of common English: that what it says about inductive arguments is as trivial, and as well-known to normal English-speakers, as what "A father is a male parent" says about fathers. These suppositions are not only false, but grotesque, and the exact opposite of the truth. The simplest way to prove this is to show that we have ample testimony, from authorities which are numerous, recent, high, and even in a sense irresistible, to the *falsity* of M.

In the first place *everyone,* in his bones, nerves, and muscles, believes that M is false. Strike a match and look at the flame. Then try not to believe that you would feel heat if you held your hand an inch over it. You cannot do it. You cannot even be *less* confident, about this future thermal phenomenon, than you are about the present visual phenomenon of the flame. This is an example, of course, of Hume's favourite kind of inductive inference, and the kind in relation to which his entire argument for scepticism was in fact conducted: what he calls "the inference from an impression to an (associated) idea"[21], after we have had "a long course of uniform experience"[22] of the conjunction of two properties, such as being a flame and being hot. The corresponding inference *before* experience, or (as Hume likes to say) in *Adam's* situation, is of course the inference just from "This is a flame" to "This will be hot". Now no one, as Hume is always saying, takes *that* premise as a reason to believe that conclusion, and still less would anyone mistake it for an *absolutely conclusive* reason to believe it. But then, as Hume is also always saying, once experience *has* supplied us with the additional premise that all the many flames observed so far have been hot, we *do* draw the conclusion that a flame as yet untested will likewise be hot; and draw it, with a degree of confidence which is introspectively indistinguishable from that with which we conclude, given that all men are mortal and Socrates is a man, that Socrates is mortal. In other words, we all do believe that, contrary to what M says, the observation-statement about past flames is sufficient, as an additional premise, to turn the argument from "This is a flame" to "This will be hot" into a valid one. At least, our bones, nerves and muscles believe so.

These authorities against M will be admitted to be numerous and recent, and even in a sense irresistible. But they may be thought to be rather low. So let us turn to Bacon and Mill, who are sufficiently high authorities on induction. And let us ask what they would have thought of the following inductive argument. "The canary was alive and well when we left the room an hour ago; but it is dead now. Gas from the oven was leaking into the room during that time. So, if nothing else caused the canary's death, the gas did."

This is, of course, a homely example of the very kind of inductive argument with which Bacon and Mill were especially occupied: "eliminative induction", as Mill aptly called it. The argument is invalid, just as Hume's premise G requires. The validator of it is the proposition that something caused the canary's death. This proposition is indeed, just as Hume's result M requires, neither a necessary truth nor a proposition about the observed. But the question is, was this fact *obvious* to Bacon and Mill? It would take a very bold man, or a very ignorant one, to say so.

From what Bacon wrote about inductive arguments of essentially this kind, it seems never to have crossed his mind what kind of proposition its validator might be: for the simple reason that he seems to have thought such an argument valid as it stands. Mill at least knew better than that, and accordingly he tried for a while, in Book III of his *Logic,* to show that the validator of such an argument is, after all, known from experience; or rather (with his characteristic rigour) that anyway it nearly is; or that it is known from experience, at any rate with "all the assurance we require for the guidance of our conduct".[23] ('Conduct'?!) But this apparent modesty Mill was unable to sustain: his real confidence in what he called the Law of Universal Causation was too deep. To suppose that the deterministic assumption (that the canary's death had a cause) was not available to inductive reasoners, in 1843, evidently seemed to Mill merely a solemn farce, and he could not keep it up. So in the end he simply throws up in impatience the question of the validity, or the curable invalidity, of eliminative induction. By doing so he seems, to 20th-century philosophers, as he would have seemed to Hume, to have left his philosophy of science in ruins. In his own century, however, there were very few who were of that opinion.

But leave even these mighty dead out of it. Consider the argument about the canary, and let us ask ourselves this. What *experimental scientist, now,* would have any more patience than Mill had, with someone who tormented him with reminders that the additional premise, which this argument needs to be valid, is not known to be true either *a priori* or from experience? Come to that, how many experimental scientists would be conscious, any more than Bacon was, that the argument is not valid as it stands? These questions answer themselves.

So very wide of the truth, then, is the belief that M is a truth which has always been obvious, at least to philosophers. As for the suggestion which some philosophers have made in recent decades, that M is known to every competent English speaker, like "A father is a male parent" ... I blush for my profession! Quite the contrary to all this, to bring to light the truth M about inductive arguments, required the peculiarly fixed, strong, and passionless gaze by which Hume was distinguished in mind as he was in body. *After* Hume, of course, you do not need to be a genius to know that M is true; but that is a little different.

It is not only dead philosophers or living scientists, however, who have not fully taken in the truth of M, or have not perceived the full extent of its consequences for empiricist philosophy of science. The same is true of most empiricist philosophers now.

The grounds which Hume *explicitly* gave for C amounted, as we saw, only to M. Yet C is his shocking conclusion about induction; while M is so far from being shocking, at least to philosophers now, that the difficulty with it is rather, as we have seen, to secure recognition of its importance. Hume proceeded, in other words, as though his premises yield a result which is even stronger and more important than M.

They do, too. What this result is, will become clear if we ask ourselves the following natural question. Why did Hume consider, as candidates for the position of available validators of induction, *necessary truths, observation-statements, and no others?* The answer is obvious. To be available to inductive reasoners, a validator of their arguments must at least be such that it can be part of a reason to believe the conclusions of induction; and Hume thinks that only necessary truths and observation-statements can be part of a reason to believe anything. In other words, Hume was here drawing again on two of the premises of his earlier argument for empiricism. He was taking M not on its own, but in conjunction with E and F. And when that is done, a result which is even more important than M does follow.

It follows at once from M, E, and F, that the validator R of an inductive argument from P to Q is not a reason or part of a reason to believe the conclusion Q. And necessarily, if even the *weakest* validator R of an argument from P to Q is not a reason or part of a reason to believe Q, then *a fortiori* any stronger validator of the argument cannot be a reason or part of a reason to believe Q either. For any stronger validator than R will be logically equivalent to R–and–S for some S, and if R cannot be even part of a reason to believe Q, then evidently no proposition logically equivalent to R–and–S can be so either. That is, no validator of an inductive argument can be even part of a reason to believe the conclusion of that argument. Taken with E and F, then, M entails

N: Any inductive argument is invalid, and any validator of it is not a reason or part of a reason to believe its conclusion.

This is an enormously important result of Hume's argument. I call it the thesis of the incurable invalidity of induction. Some invalid arguments, we have seen (for example the one about the bird on the post) are only *curably* invalid; a validator of them is available to the arguer, at least in the sense that such a validator can be part of a reason to believe the conclusion of the argument. What N says is that inductive arguments are *not* like that: for *their* invalidity, no cure is available. Any additional premise, if it *is* sufficient to make the conclusion of an inductive argument logically follow, is *not* a reason or part of a reason to believe that conclusion. In other words, the fallibility or invalidity of inductive arguments (the possibility of their having a false conclusion even though their premises be true) is a feature absolutely inseparable from them.

Whereas many philosophers now need to be reminded of the importance of *M,* the importance of the present result *N* is obvious to them all. Indeed, many philosophers, beginning with Hume himself, believe that, or at least proceed as though, *N* is so devastating a result about induction that the sceptical conclusion *C* follows from it at once. If induction really is, as *N* says it is, not only invalid but incurably so, does it not follow that induction is unreasonable, as *C* says it is?

Good philosophers have a very exacting standard of what constitutes reasonable argument; and other things being equal, one philosopher is better than another, the *more* exacting his standard of reasonable argument is. The highest possible such standard would say, that the premise of an argument is not a reason to believe the conclusion, unless the argument is actually valid; and it is, accordingly, to such a standard as this that all good philosophers more or less incline. They have the deepest reluctance, consequently, to admit that an argument can be a reasonable one, if it is not only invalid, but cannot be turned into a valid argument by any additional premise which can form part of a reason to believe its conclusion. Hence the admission of *N,* that *inductive* arguments are all in this position, is bound to impose at least some strain on any good philosopher's belief in the reasonableness of induction. It is natural, therefore, for a good philosopher to think that *C* follows from *N.* He will even, other things being equal, move from *N* to *C* the more easily, the better philosopher he is. There is nothing at all surprising, then, but quite the reverse, in Hume and many other philosophers having proceeded as though *N* entails *C.*

Nevertheless, some other philosophers (of whom I am one) resist this step from *N* to *C.* We have a less exacting standard of reasonable argument than most philosophers incline to. We say that an argument can be invalid, and even incurably so, and still its premise be a reason to believe its conclusion. It *is* so, we say, with some inductive arguments in particular. Hume's result N we accept, and we admire it, as a profound truth about induction which his argument brought to light. But the sceptical conclusion C which Hume drew from N does not follow, we say, and is false.

This kind of philosopher, the 'inductive probabilist' as he may be called, does not think of the invalidity of inductive arguments as a mere surface blemish of them. He knows better than that, for he has taken Hume's result N to heart, and therefore holds that the invalidity of induction is incurable. Still less will he join those philosophers who *search for* a validator of induction. Invalidity which cannot be cured, he considers, had better be endured. On the other hand he still maintains that some inductive arguments are reasonable, in the sense in which C says that none are; that is, he maintains that their premises are a reason to believe their conclusions. Of course he does not regard the reasonableness of those arguments as an *extrinsic* feature of them: as consisting in the fact that some *other* argument, which has the same conclusion and augmented premises, is actually valid. On the contrary, he regards the reasonableness of those inductive arguments which are reasonable as an intrinsic logical feature of them; just as, for example, their invalidity is. So while he admits that inductive arguments have an incurable infirmity, in that it is possible for their premises to be true and conclusion false, he does not obsessively concentrate on this logical feature of them to the exclusion of every other.[24]

But most philosophers, it must be admitted, consider the inductive probabilists' position a feeble evasion, and one impossible to maintain. Many suspect that the inductive probabilist, despite the lip-service he pays to Hume, has never really taken in the full force of his argument. Some even suspect that he is engaged, most embarrassingly, in defending a position about induction which Hume himself had already shown, in the course of the very argument we are discussing, to be indefensible.

'Consider' (these critics say) 'an inductive argument, for example that from P "All the many flames observed in the past have been hot", to Q "Tomorrow's flames will be hot". There is no connection whatever between the premise and the conclusion. P and Q are propositions entirely logically independent of one another. Nevertheless, you tell us, P is a reason to believe Q. Now, is it not obvious that, if this *is* so, it is because there is some *connection* between past flames and tomorrow's flames, or between being a flame and being hot, or between the observed and the unobserved? To make P a reason to believe Q, there must be *some* ground in nature, some fact about the cosmos, some "cement of the universe" (in Hume's phrase) which, taken along with P, *logically* connects that premise with the conclusion Q; that is, turns the original inductive argument into a valid one. Yet you reject Hume's *C* while you accept his *N*. There is no proposition, you say, which is at once part of a reason to believe Q, and sufficient to make the argument from P to Q valid. If so, then *a fortiori* there is no true proposition of that kind. And what is this but to say that there is in the nature of things no ground for inferring Q from P, or that P is *not* a reason to believe Q? Forswear these evasions, then, and admit at any rate the truth of Hume's conditional, that if *N* is true, *C* is: that if induction really is incurably invalid, then it is unreasonable. Or, of you persist in affirming *N* and denying C, at least tell us what it is, according to you, which *makes it true* that (for example) P is a reason to believe Q. It cannot be, that the argument from P to Q is valid, or only curably invalid; for you accept *N,* and insist that the argument is not so. What *is* it, then, that makes P a reason to believe Q? If you tell me that it is just an ultimate fact of inductive logic, or of the theory of probability, that P is a reason to believe Q, then I will know what to think of your so-called inductive logic: that it is simply speculative metaphysics in disguise.'

Criticism of this kind has often been thought to be fatal to inductive probabilism. Its plausibility must have been felt, at least at times, even by the inductive probabilist himself. It can fairly be summed up thus: "If the universe were not connected or cemented in some way (that is, if there were no true validator of at least some inductive arguments), then Hume's scepticism about induction would be true."

Some of these critics of inductive probabilism have an *anti*-sceptical intent. They intend to go on to say that, since Hume's inductive scepticism *C* is plainly false, the universe must in fact be cemented or connected in some way. (As to the nature of the cement, they may and do differ. Some of them say it is causation; others that it is a certain connection which exists between properties; others again that it is the providence of God; etc.) A second group of the critics of inductive probabilism have a sceptical intent. They mean to go on to say that, since the universe is in fact *not* cemented or connected, Hume's inductive scepticism *C* is *true*. Of this second group a recent example is Popper, in *The Logic of Scientific Discovery*.[25] Of the first, a recent example is D. M. Armstrong.[26]

For our purposes, however, the difference between these two groups of critics does not matter. What matters is what they agree on. For this can be shown to be a complete mistake.

What the critics of inductive probabilism unite in believing is that Hume's inductive scepticism would be true if the universe were not cemented or connected in some way; or what is equivalent, that his inductive scepticism would be false only if the universe were cemented or connected. That the universe is cemented or connected, whatever exactly it means, is a proposition which is, as an additional premise, sufficient to turn at least some inductive arguments into valid ones. Inductive arguments, however, as well as being invalid, all have contingent conclusions; and any additional premise, which is sufficient to turn an invalid argument with a contingent conclusion into a valid one, must be contingent itself. It is therefore a contingent proposition that the universe is cemented or connected. Since the negation of any contingent proposition is itself contingent, it is also a contingent proposition that the universe is *not* cemented or connected. Our critics therefore all imply that Hume's scepticism about induction would be true if a certain contingent proposition ("no cement") is true, false only if a certain contingent proposition ("cement") is true. But a proposition which is true on a certain contingent condition and false otherwise, is contingent itself. All these critics therefore imply that Hume's inductive scepticism is a contingent proposition.

But this is so extreme a misconception of the nature of Hume's *C* that no one, I believe, will venture to defend it, once it is thus explicitly stated.

On its very face it is most implausible. Recall the proposition *C*, or any of Hume's own words for it: for example, "even after the observation of the frequent or constant conjunction of objects, we have no reason to draw any inference concerning any object beyond those of which we have had experience."[27] This certainly does not *appear* to be a contingent claim about the overall character of the actual universe. It appears to be, rather, a *logical* thesis of some kind: a proposition about whether certain propositions are or are not a reason to believe certain other propositions.

If *C is* contingent, then Hume's argument for it must either have been invalid, or have had at least one contingent premise. For remember, 'there can be no *demonstrative* arguments for a matter of fact and existence'! , That Hume's argument for C was valid, we are assuming. Where, then, is its contingent premise? The reader has only to recall E, F, G, H, and J, to see that there is no contingent premise in Hume's argument for C, according to my account of it above. Nor is there a contingent premise in Hume's argument for C, according to any other account of it which is worthy of consideration in other respects. Such a premise, then, is not easy to find in Hume's argument for inductive scepticism. Yet if it is there at all, it must be a *huge* one; since it has to entail a conclusion, C, which on the present hypothesis is nothing less than a contingent proposition about the entire universe. We must ask to have this elusive giant—this Yeti, as it were, among Hume's premises—pointed out to us. But it is not really necessary to rely here on the method of challenge. Any hope, or fear, that the challenge just issued might be met, can be very easily extinguished. For suppose that Hume's inductive scepticism C *were* contingent. Then, since what it says, it says about *any* inductive argument, it would be a *universal* proposition, as well as a contingent one. Contingent universal propositions, however, are a species of *contingent propositions about the unobserved;* and we know what Hume here implies about all of *them*. His empiricism B says that only propositions about the observed can be a reason to believe them; and his inductive scepticism C says that even propositions about the observed are no reason to believe them. On the hypothesis that C itself is contingent, then, Hume's scepticism A about the unobserved, which follows from B and C, would imply, concerning inductive scepticism C itself, that there is no reason to believe it; any more than to believe, say, that tomorrow's flames will be hot.

Well! Philosophically, of course, no result could possibly

be more welcome than this, to any empiricist who denies inductive scepticism: if only, as a matter of Hume-exegesis, he could believe it! 'Hume's inductive scepticism, given his empiricism, entails that there is no reason to believe his inductive scepticism': what an optimum result! Alas, as a matter of the exegesis of Hume, even the inductive probabilist finds this far too good to be true. Still less will any other student of Hume be able to believe it. Hume's inductive scepticism C, however it is to be refuted, if indeed it can be refuted at all, is certainly not a proposition which makes its own refutation unnecessary, by committing suicide at birth in this obliging and even graceful manner. It *would* be a proposition of exactly that kind, however, if it were contingent. Therefore it is not contingent.

It is not hard to see how the anti-sceptical critics of inductive probabilism have been led into the exegetical absurdity just noticed. They have not fully taken in Hume's result N. No one would search for a cement of the universe which would validate inductive arguments, if he were once fully persuaded that anything which *was* equal to that task would not be part of any reason to believe the conclusions of induction. The position of the *sceptical* critics of inductive probabilism, such as Popper, is much less intelligible. They are empiricists, and even inductive sceptics, yet somehow they have had revealed to them what they imply is a natural law: that the cement-content of the universe is constant zero. Criticism is superfluous in such a case.

Both groups of critics suppose, as I said earlier, that *they* adhere more rigorously than the inductive probabilist does to the truths about induction which Hume's argument can teach us. In fact, as we have now seen, the boot is on the other foot. The inductive probabilist has taken in, far better than his anti-sceptical critics, the truth of Hume's result N. And he has taken in, far better than any of his critics, the non-contingent character of Hume's conclusion *C*. As for the belief that the inductive probabilist is reviving a position refuted in advance by Hume himself, this is a mere myth. Its only foundation is ignorance of the texts; ignorance, in particular, of what Hume meant by the phrase "probable arguments". Far from having refuted inductive probabilism, Hume never so much as considered it. He scarcely could have done so, because inductive probabilism came into being, in the modern period of philosophy, or at least assumed a definite form, only in *response* to his sceptical attack on induction.[28] Hume's only question about induction was, as I said earlier: what can *validate* it (while being also part of a reason to believe its conclusions)? Finding, *N*, that nothing can, he forthwith concluded, *C*, that induction is unreasonable. That inductive arguments, or that any arguments, might be reasonable *although* incurably invalid, is a position which Hume nowhere attempted to exclude.[29]

The inductive probabilist can easily show, too, that his critics' implied philosophy of logic is no more satisfactory than their implied interpretation of Hume.

'You challenged me to say' (he might reply to his critics) 'what makes it true that "All the many flames observed in the past have been hot", is a reason to believe "Tomorrow's flames will be hot"; or to say, in general, what makes it true that Hume's inductive scepticism *C* is false. And in accordance with your misconception of the nature of *C,* you were then demanding a *contingent* truth-maker for this assertion of mine. But inductive scepticism *C* is not a contingent proposition. No more, then, is my denial of it contingent. The negation of *C* is, like *C* itself, a logical thesis in a broad sense, and whichever one of the two is true, that proposition does not require a contingent truth-maker, any more than other propositions of the same kind do.'

'When we say' (he might continue), 'as we all do say, and as Hume's *D* implies, that a tautology, for example, is not a reason to believe "Tomorrow's flames will be hot", does *this* assertion of ours require a ground in nature to make it true? When we say, as all philosophers do, that "All men are mortal and Socrates is mortal" is not an absolutely conclusive reason to believe "Socrates is a man", does this assertion depend for its truth on some cosmic contingency? If it does, what is that contingent feature of the universe which makes undistributed middle a fallacy? Is it an unfortunate local deficiency of cement, perhaps, or a vein of actual anti-cement which runs through our fallen world? When philosophers say, as almost all of them do say, that "All men are mortal and Socrates is a man" is an absolutely conclusive reason to believe "Socrates is mortal", does their assertion require a contingent truthmaker? If it does, then deductive logic too, no less than non-deductive logic, will be 'speculative metaphysics'. To every one of these questions, the answer is obviously "no". And no more does my assertion, when I say that some propositions about the observed are a reason to believe some contingent propositions about the unobserved, require any contingent fact to make it true.'

There is one compromise, however, which the inductive probabilist can and should offer to his critics. He should undertake to reveal what the contingent fact is, which makes the premises of some inductive arguments *a reason* to believe their conclusions, on the very day that his critics reveal what the contingent fact is, which makes the premise of every inductive argument *not an absolutely conclusive reason* to believe its conclusion. Mutual disclosure of all contingent assets is a fair principle. Let the ground in nature of the reasonableness of some inductions be disclosed, then, in return for disclosure of the ground in nature of the invalidity of all of them.

Nor is the inductive probabilist obliged to maintain,

wherever the premise of an induction *is* a reason to believe the conclusion, that this is an *ultimate* logical feature of those arguments. It may sometimes be possible to show that it is a *derivative* one. It may be possible, that is, to derive the conclusion that certain inductive arguments are reasonable, from premises about the reasonableness of certain *non*inductive arguments. Indeed, it is already known (thanks originally to Bernoulli and Laplace[30]) that this can in certain cases be done. But any premise of such a derivation will be, like the conclusion of it, a proposition not of a contingent but of a logical kind. And even where such a derivation is possible, the reasonableness of the inductive arguments in question remains an intrinsic feature of them, even though not an ultimate one.

The philosophical dispute between inductive probabilism and its critics, as I have presented it, arose from an historical dispute, about the interpretation of Hume's argument for inductive scepticism. The question was whether, supposing induction to be incurably invalid, as *N* says it is, it follows that induction is unreasonable, as *C* says. The inductive probabilist believes it does not follow. Hume, and the critics of inductive probabilism, believe that it does. But now, there is nothing to prevent us from condensing this whole cloud of philosophy and of Hume-exegesis into a single drop of elementary logic. Does *C* follow from *N*, or does it not?

Well, *C* says this: that the premise of an inductive argument is not a reason to believe its conclusion. *N* says this and only this: that any inductive argument is invalid, and that no validator of it is a reason or part of a reason to believe its conclusion. But evidently, from the fact that no *validator* R, of an inductive argument from P to Q, is a reason or part of a reason to believe Q, it does not follow that the *premise P itself* is not a reason to believe Q. Yet that is what *C* says. So *C* does not follow from *N*. The incurable invalidity of induction is no proof of its unreasonableness.

THERE is, therefore, a gap in Hume's argument for inductive scepticism C. Proceeding, as we are, on the assumption that the argument which Hume intended was valid, we therefore have no alternative but to suppose that his argument had some premise which he did not state: some tacit assumption which, when it is added to his other premises, is sufficient (and of course no more than is necessary) to turn his argument for C into a valid one. For our purposes it is necessary to identify all the premises of his argument. We need, therefore, to identify this suppressed premise: the *validator* (for it is nothing less) of Hume's own argument for scepticism about induction.

The *general* nature of this missing premise is obvious enough. One who believes that C follows from N regards it as necessary, in order for the premise of an argument to be a reason be believe the conclusion, that that argument be valid, or at least not incurably invalid. One who denies that C follows from N, denies that this *is* a necessary condition for an argument to be reasonable. The former philosophers therefore (as was said earlier) have a higher or more exacting standard than the latter, of what it is for the premise of an argument to be a reason to believe its conclusion. As we called the latter, with obvious propriety, 'inductive probabilists', so we may call the former 'deductivists'. For their standard of a reasonable argument, whatever exactly it may be, is one which demands that, if P is to be a reason to believe Q, then Q is *deducible* either from P itself, or from P along with such limited additional premises as can be themselves part of a reason to believe Q. Hume, since he believes that C follows from N, is one of these deductivist philosophers. We are therefore entitled to call the unexpressed assumption, which enabled him to mistake the argument from N to C for a valid one, Hume's *deductivist* premise; even though it is yet to be determined exactly what it says.

The presence of such an assumption in Hume's argument, as it is obvious, has often been noticed. Many writers have detected, not only in his argument for inductive scepticism, but elsewhere in Hume's philosophy, the influence of an inexplicit standard, of a 'high' or 'deductivist' or 'rationalist' kind, as to what constitutes reasonable argument. Thus one writer says, for example, that in Hume's argument for inductive scepticism, "the tacit assumption [was that] all rational inference is deductive".[31] Another says that Hume's assumption was, that "arguments are deductive or defective"[32]. Many other writers could easily be cited to the same effect.

The last-quoted version of Hume's deductivism is too vague to be of any use to us; for the writer does not explain, and it is not obvious, what he means by "defective". The previously-quoted one is identical with the version of deductivism which, in my earlier-published account of this argument, I myself attributed to Hume. For I there concluded that Hume's unstated premise was, that an argument is reasonable only if it is valid; or in other words, that P is a reason to believe Q, only if Q is deducible from P.[33]

But it is easy to see, in the light of the more detailed account of Hume's argument which has been given here, that this identification is wrong. This simple version of deductivism makes Hume assume both too much and too little. Too little, because it takes no notice at all of the distinction between arguments which are (simply) invalid, and arguments which are *incurably* so. Too much, because it is, evidently, stronger than the validator of the argument from N to C. (To attribute it to Hume is therefore to attribute to him more than is needed to make his argument valid: a serious fault in exegesis.) And both of these defects, it will be obvious, arise from the same source: namely, that this version of deductivism 'engages' only with the *first* clause of N (that inductive arguments are invalid).

Hence on this identification of Hume's deductivism, the second clause of *N*, which adds that the validity of induction is incurable, plays no essential part in his argument at all.

But on the contrary, it is the second clause of *N* which Hume's entire argument had been directed to establishing; not the first clause of it, the *mere* fact that inductive arguments are invalid. *That* was assumed from the outset, in Hume's premise *G*. What Hume argued for, and argued successfully for, was *I* and *K*: that *the validator* of inductive arguments is neither necessarily true nor observational. It was *this* result, combined with his assumption, which follows from *E* and *F*, that it is *only* necessary truths and observation-statements which can be even part of a reason to believe another proposition, from which Hume validly inferred that the invalidity of induction cannot be cured at all; that is, *N*. And some deductivist assumption, the exact nature of which we wish to identify, conjoined with *N* and perhaps with some other premises of his argument, then carried Hume to inductive scepticism *C*.

With this recapitulation of Hume's argument before us, then, let us ask afresh, what is the assumption, of a deductivist kind, which is implicit in the argument?

The conclusion to be reached, *C*, says that the premise of an inductive argument is not *a reason to believe* its conclusion. We ought, therefore, in trying to identify the missing deductivist premise of the argument for this conclusion, to take account not only of *N*, but of anything which the *already-identified* premises of the argument say, about what is required in order for one proposition to be a reason to believe another. *E* and *F* contain everything that there is of that kind, in the premises of the argument which have been already identified. We ought therefore, in trying to identify the deductivist assumption which supervened between *N* and *C*, to take *N* not on its own, but in conjunction with *E* and *F*.

Once this is done, the missing premise stands out clearly. The most that is entailed about inductive arguments by the conjunction of N, E, and F, is

M^+:   Any inductive argument is invalid, and any validator of it is neither a necessary truth nor a proposition about the observed.

(I call this M^+ because it says, about any validator of inductive arguments, just what M says about the weakest one.) The missing deductivist premise is therefore the validator of the argument from M^+ to C. That proposition is evidently

O: P is a reason to believe Q only if the argument from P to Q is valid, or there is a validator of it which is either a necessary truth or a proposition about the observed.

This proposition, therefore, is Hume's deductivist premise.

In order to satisfy ourselves of this, it is sufficient to cast our minds back over Hume's argument, and ask ourselves the following simple question. What, after all, did Hume have *against* inductive arguments? What *is* it, at bottom, about inductive arguments, which elicits from him the dire verdict C against them? Well, certainly not the mere fact that they are not valid. *That* feature of them was taken for granted by Hume (in premise G) from the start, and is, besides, a feature, unimportant in itself, which inductive arguments share with many arguments that neither Hume nor anyone else condemns; for example, the argument mentioned earlier, about the bird on the post. No, what makes inductive arguments unreasonable in Hume's eyes is that, precisely *unlike* the argument about the bird on the post, *their* invalidity cannot be cured by any additional premise which might be supplied either by *a priori* knowledge or by experience. *That* is what Hume has against induction. Which is to say that he assumed that, if an argument is a reasonable one, it is either valid, or can be made so by an additional premise which is either necessarily true or observational. Which is to say that his tacit premise of a deductivist kind was O; and that what that premise engaged with was M^+.

In the presence of E and F, N entails M^+; but equally, in the presence of E and F, M^+ entails N. In other words, given Hume's assumptions, that a reason to believe another proposition must be directly accessible to knowledge or reasonable belief, and that all and only observation-statements and necessary truths are so directly accessible, M^+ and N are logically equivalent. From a logical point of view, therefore, it is immaterial whether we regard deductivism O as engaging with N or with M^+; that is, whether we regard the last step of Hume's argument for C as having the structure

$$\left.\begin{array}{l} E \\ F \\ \\ N \\ O \end{array}\right\} \to C$$

or the structure

$$\left.\begin{array}{l} E \\ F \\ N \end{array}\right\} \to M^+ \left.\begin{array}{l} \\ \\ O \end{array}\right\} \to C$$

But for the reason given in the preceding paragraph, the latter interpretation is to be preferred.

The structure, then, of Hume's argument for inductive scepticism was the following:

$$
\left.\begin{array}{c}H\\ J\end{array}\right\}\to I
\quad
\left.\begin{array}{c}\left.\begin{array}{c}H\\ G\end{array}\right\}\to L\to K\end{array}\right\}\to M
\quad
\left.\begin{array}{c}E\\ F\end{array}\right\}\to M
\quad
\left.\begin{array}{c}E\\ F\end{array}\right\}\to N
\quad
\left.\begin{array}{c}E\\ F\end{array}\right\}\to M^{+}
\quad
\left.\begin{array}{c}\phantom{M^{+}}\\ O\end{array}\right\}\to C
$$

It is important to realise, as was pointed out earlier, that it is possible consistently to be an inductive sceptic without being a sceptic about the unobserved; that is, that A does not follow from C alone, but only from C conjoined with empiricism B. In the same way it is important to realise that it is possible consistently to be a deductivist, without being an inductive sceptic; that is, that C does not follow from O alone, but only from O conjoined with M, which says that no validator of induction is either necessarily true or observational.

It ought to be obvious that C does not follow from O. For deductivism O says nothing about inductive arguments, or indeed about any particular class of arguments, at all. All it does is to allege that a certain condition is necessary in order for an argument to be a reasonable one. Clearly, this on its own cannot entail that some particular class of arguments satisfies, or fails to satisfy, that condition. It is only when O is conjoined with M^{+}, which is about inductive arguments, and says of them that they *fail* to satisfy the condition demanded by O for reasonable arguments, that C, the unreasonableness of induction, follows.

It ought to be equally obvious that C does not entail O, either. C says only that inductive arguments are not reasonable.

From such a proposition as that, it is evidently impossible to deduce any positive condition which arguments in general must satisfy in order to be reasonable. But such a condition is precisely what O lays down.

Deductivism O and inductive scepticism C are, then, neither of them deducible from the other. These facts are important. For these two theses, although in fact independent, are nowadays often, or rather, usually, inextricably confused with one another. The historical reason for this confusion is, of course, just the fact that the two theses *are* closely connected in the context of Hume's argument about induction, while that argument is nowadays vividly, though confusedly, present to the minds of most philosophers. Hence nowadays the deductivist believes himself obliged to be a sceptic about induction; the friend of induction believes himself bound to reject deductivism; the inductive sceptic imagines himself bound to be a deductivist; the enemy of deductivism considers himself safe from inductive scepticism; and every single one of these beliefs is false. It is scarcely possible, in fact, to overestimate the damage which has been done, in the way of positive error but even more in the way of mere confusion, by the failure to recognise that deductivism, and scepticism about induction, are independent theses. And this damage has been inflicted, not only on our ability to understand Hume's argument about induction, but also, and even more importantly, on what is based on that argument: twentieth century philosophy of science.

It is the mistaken belief that O entails C, rather than the converse belief, which has been the more productive of error and confusion. This belief has had the effect, among others, of making unintelligible to most philosophers nowadays a philosophy of science which is as recent, as influential, and as intelligible, as that of J. S. Mill. For the kind of position which was pointed out (three paragraphs back) as a logically possible one, namely deductivism without inductive scepticism, was in fact that of Mill. What Mill really believed, as was indicated earlier, is that M^+ (and even M) is *false*. That is, he thought that the deterministic validator of eliminative inductions (such as the argument about the dead canary) is *observational*. *That* is why he could consistently be, what he was, a deductivist and yet no sceptic about induction.

3

Now, among the premises of Hume's argument for inductive scepticism C, which one is the key to the scepticism or irrationalism of that conclusion? What is that premise, without which this argument would have neither C nor any other irrationalist thesis as a consequence?

Well, the only premises of the argument for C are G, H, and J, E and F, and O.

The first five of these premises, however, entail nothing of a sceptical or irrationalist kind about induction. They entail, indeed, M, that the invalidity of induction is incurable by any observation-statement or necessary truth. And they entail, what is equivalent to M^+ in the presence of E and F, N, that the invalidity of any inductive argument is incurable by any additional premise which is even part of a reason to believe its conclusion. But these results M^+ and N are the most that these five premises entail about induction. And there is nothing sceptical or irrationalist about either of them. They do not entail C. (They say, indeed, no more than is acknowledged nowadays by almost all philosophers, and not just by inductive sceptics: that induction 'cannot be turned into deduction'.) Thus before the sole remaining premise, deductivism O, comes into the argument, there is nothing to necessitate C or any other irrationalist conclusion about induction. Once that assumption does come in, however, it engages with M^+, and scepticism C concerning induction is an inevitable result. It is, therefore, deductivism which is the key premise of Hume's argument for inductive scepticism.

Nothing fatal to empiricist philosophy of science, in other words, follows from the admission that arguments from the observed to the unobserved are *not the best;* unless this admission is combined, as it was combined by Hume, with the fatal assumption that *only the best will do.*

4

FINALLY, it is worthwhile, for the sake of getting an overall view of Hume's argument for scepticism about the unobserved A, to put together the two parts of its structure-diagram: the argument for B, and the argument for C. This is done below. For convenience of reference, all the elements of the argument are also listed below: and I have here given each of them a summary title, which may be found helpful.

$$
\left.\begin{array}{c}\left.\begin{array}{c}H \\ J \\ H \\ G\end{array}\right\} \begin{array}{c}\rightarrow \rightarrow I \\ \rightarrow L \rightarrow K\end{array}\right\} \rightarrow M \quad \left.\begin{array}{c}\left.\begin{array}{c}E \\ F\end{array}\right\} \rightarrow N \\ O \\ D \\ E \\ F\end{array}\right. \left.\begin{array}{c}\left.\begin{array}{c}E \\ F\end{array}\right\} \rightarrow M^+ \\ \rightarrow B\end{array}\right\} \rightarrow C \\ \rightarrow B
$$

A (Scepticism about the unobserved) There is no reason to believe any contingent proposition about the unobserved.

B (Empiricism) Any reason to believe a contingent proposition about the unobserved is a proposition about the observed.

C (Inductive Scepticism) No proposition about the observed is a reason to believe a contingent proposition about the unobserved.

D (Impotence of the a priori) No necessary truth is a reason to believe any contingent proposition.

E (Accessibles necessary or observational) A proposition is directly accessible to knowledge or reasonable belief if and only if it is either a necessary truth or a proposition about the observed.

F (Reasons must be accessible) If P is a reason or part of a reason to believe Q then P is directly accessible to knowledge or reasonable belief.

G (Induction invalid without Resemblance) Any inductive argument is invalid, and the validator of it is a Resemblance Thesis.

H (Resemblance a contingent feature of the universe) A Resemblance Thesis is a contingent proposition about the unobserved.

I (Resemblance not provable a priori) A Resemblance Thesis is not deducible from necessary truths.

J (No contingents provable a priori) No contingent proposition is deducible from necessary truths.

K (Resemblance not provable a posteriori) A Resemblance Thesis is not deducible from propositions about the observed.

L (Induction to Resemblance circular if valid) A Resemblance Thesis is deducible from propositions about the observed only when to the latter is conjoined a Resemblance Thesis.

M (The validator of induction not necessary or observational) Any inductive argument is invalid, and the validator of it is neither a necessary truth nor a proposition about the observed.

M⁺ (No validator of induction necessary or observational) Any inductive argument is invalid, and any validator of it is neither a necessary truth nor a proposition about the observed.

N (Invalidity of induction incurable) Any inductive argument is invalid, and any validator of it is not a reason or part of a reason to believe its conclusion.

O (Deductivism) P is a reason to believe Q only if the argument from P to Q is valid, or there is a validator of it which is either a necessary truth or a proposition about the observed.

CHAPTER FIVE

FURTHER EVIDENCE FOR THIS IDENTIFICATION

ECAPITULATING the conclusions of Chapter Four, but omitting now the details of the sub-arguments for empiricism and for the fallibility of induction, Hume's argument was the following:

$$
\left. \begin{array}{l} \text{Falibility of} \\ \text{Induction,} \\ \text{Deductivism} \end{array} \right\} \rightarrow \text{Inductive Scepticism} \left. \begin{array}{l} \\ \end{array} \right\} \begin{array}{l} \text{Scepticism} \\ \rightarrow \text{about the} \\ \text{Unobserved} \end{array}
$$
$$
\text{Empiricism} \left. \begin{array}{l} \end{array} \right\}
$$

That is, the immediate grounds of his scepticism about the unobserved were the thesis that only propositions about the observed can be a reason to believe anything about the unobserved, plus the thesis that even they are not such a reason. And for this latter sceptical thesis, his ultimate grounds were premises which entail that the invalidity of inductive arguments is incurable by any additional premises which are either observational or necessarily true; plus the assumption that the premise of an argument is no reason to believe its conclusion, unless the argument is valid, or can be made so by additional premises which are either observation-statements or necessary truths. But from empiricism, or from inductive falibilism, or from their conjunction, no sceptical or irrationalist consequence follows. When they are combined with deductivism, however, first scepticism about induction follows, and then scepticism concerning any contingent proposition about the unobserved. The key premise of Hume's argument, therefore, in the sense of being that premise without which the argument would have no sceptical or irrationalist consequences, is deductivism.

All of this is exactly true of Popper as well. His irrationalism about scientific theories is no other than Hume's scepticism concerning contingent propositions about the unobserved; nor are his grounds for it other than Hume's. Popper is no less an empiricist than Hume: he does not believe, any more than Hume did, that any propositions except observation-statements can be a reason to believe a scientific theory. And at the same time he is, as he is always telling us, a Humean sceptic about arguments from the observed to the unobserved. For this inductive scepticism in its turn, Popper's argument is just, as he tells us, that 'flawless gem' of an argument which was Hume's: from the fact that inductive arguments are invalid, and that this condition cannot be cured by additional premises either observational or necessarily true; plus the deductivist assumption that if an argument is of this kind, then its premise is no reason to believe its conclusion. In Popper's philosophy of science, therefore, as in Hume's, the premise on which all the irrationalist consequences depend is deductivism. And since our other authors' philosophy of science is derived almost entirely from Popper's, deductivism is the key to their irrationalism too.

Recent irrationalist philosophy of science is therefore to be ascribed (insofar as it can be ascribed to intellectual causes at all) to acceptance of the thesis of deductivism. What has been decisive in leading these authors to conclude that there can be no reasonable belief in a scientific theory, and *a fortiori* that there has been no accumulation of knowledge in the last few centuries, is a certain extreme belief, by which their minds are dominated, about what is required for one proposition to be a reason to believe another.

The *truth* of the key premise of our authors' philosophy is not (as was said at the beginning of this Part) a question with which this book is concerned. But we have now at least *identified* the proposition to which criticism of their philosophy, if it is not to be entirely indecisive, needs to be directed. To criticise our authors on the basis of the history of science, for example, is sure to be in practice indecisive at best, but is futile even in principle. For the assumption on which everything distinctive of their philosophy rests is in fact one which has nothing at all to do with science, and least of all with the history of science. It is a simple thesis in the philosophy of *logic* or of *reasonable inference,* and it is nothing more. It has not even, it should be emphasised, any necessary connection with the subject of *inductive* inference; for, as was pointed out near the end of the preceding chapter, deductivism is a thesis logically independent of inductive scepticism.

Deductivism is not, of course, *explicit* in Popper's writings; though it is more nearly so there than it is with Hume. At the same time I know of no reason to doubt that Popper would accept the attribution of that thesis to him. Of course virtually no philosopher nowadays, if he were to embrace deductivism explicitly, would bother to retain that part of it which refers to *necessary truths* as possible validators of arguments. On the contrary, philosophers now assume that the addition of a necessary truth to the premises of an invalid argument will never turn it into a valid one. The reason is, that the conjunction of any necessarily true R with any P is logically equivalent just to P itself, and that two arguments cannot differ in logical value, and hence one of them cannot be valid and the other not, if they have logically equivalent premises and the same conclusion. Accordingly I too will henceforth omit the phrase "...or necessary truths" from the thesis of deductivism; and when I sometimes in the following pages call an argument *incurably invalid,* I will here mean just that no additional premise which is *observational* would turn it into a valid argument.

I have now given my answer to the question to which Part Two of this book is addressed: how did our authors come to embrace irrationalist philosophy of science? My answer is, through embracing deductivism. My main grounds for thinking this answer correct are those which have now been given: that our authors' irrationalism about science is derived from Hume, and that the key premise of Hume's irrationalist philosophy of science is deductivism.

But there are other grounds as well for thinking this answer correct. These are, in sum, that it explains extremely well a number of prominent and distinctive features of our authors' writings, including some features which seem at first quite unconnected with deductivism, or even opposed to it. This is what I now intend to show.

2

FIRST, the deductivism of our authors is what ultimately necessitated those two devices which were the subject of Part One above, and which are these authors' literary hallmark.

If you are a deductivist, then you cannot allow yourself to use, in earnest, the word, "confirms", or any of the weak or non-deductive-logical expressions. To say of an observation-statement O that it confirms a scientific theory T, entails that those two propositions stand in some logical relation such that

O is a reason to believe *T*. But this cannot be so if deductivism is true, in view of the truth of the fallibilist thesis, that neither *O*, nor *O* conjoined with any other observation-statements, entails *T*. So instead of saying that *O* confirms *T*, a deductivist, at least if he is resolved, as our authors are, not to be openly and constantly irrationalist about science, must often write that *O* 'confirms' *T*, or write that *scientists regard T* as confirmed by *O*; or write something, anyway, which, while it purports to be a statement of logic, is in fact nothing of the kind. In other words, he must often sabotage the logical expression 'confirms'.

In this way, all logical pipes, along which reasonable belief might travel from observation to scientific theories, are cut by deductivism. But our authors are also empiricists, and do not for a moment suppose that there are any sources, *other than* observation, from which reasonable belief in scientific theories might come. And neither they nor anyone else, of course, suppose that scientific theories are *directly* accessible to knowledge or reasonable belief. According to these authors, therefore, reasonable belief cannot accrue to scientific theories in any way at all. And that is why *success-words*, like "knowledge" and "discovery", which imply that reasonable belief *has* accrued to the propositions which are their objects, *must* be neutralised by our authors when they are employed in connection with science; or, if not neutralised, then simply avoided altogether.

We thus see that, of our authors' two devices for making irrationalism plausible, the primary one is the sabotage of logical expressions: the need to neutralise success-words is consequential upon that. But the sabotage of logical expressions is in its turn necessitated by a substantive philosophical thesis; and that is deductivism.

It has been widely recognised, and was admitted at the beginning of this book, that the answer in *general* terms to the question, "How have these authors made irrationalism plausible to their readers?", is: by fostering the confusion of questions of logical value with questions of historical fact, or of the philosophy with the history of science. I undertook to show *in detail* how this trick is turned. This has now been done. For, of our authors' devices for making irrationalism plausible, the basic one, we have now found, is the sabotage of logical expressions; their favourite way of doing this, we saw in Chapter Two, is by embedding a statement of logic in an epistemic context about scientists; and now, the effect of that is, precisely, to disguise historical statements as logical ones. Thus for example. the schematic logical statement, "Observation-statement O confirms theory T", attributes a certain logical value to the argument from O to T, but its ghost-logical surrogate, "Scientists regard T as confirmed by O", for all its artful suggestions of logic, is nothing but an historical proposition after all.

A deductivist philosopher of science, if he is an empiricist and an inductive fallibilist, must sabotage logical expressions which are weak. But even the strong or deductive-logical expressions will almost inevitably undergo misuse at his hands. Recall the 'primal scene' of the sabotage of a deductive-logical expression: that is, Popper contemplating the logical relation between E, "The relative frequency of males among births in human history so far is .51", and H, "The probability of a human birth being male is .9". A deductivist cannot say, what anyone else can and would say, that E is a reason to believe that H is false. For this cannot be true if deductivism is true; since neither E nor the conjunction of E with any other observation-statement entails not-H. What, then, *is* the deductivist to say about the relation between E and H? Since "falsifies" entails "is inconsistent with", he cannot, for the fallibilist reason just Mentioned, say without falsity that E falsifies H. Being cut off from using *weak* logical expressions, he cannot say, what is of course true, that E disconfirms H, or confirms not-H. Rather, then, than write down nothing at all, or something which is obviously false, the deductivist very understandably, writes instead that E 'falsifies' H; or that any scientists would *regard* E as falsifying H; or else he himself 'proposes' that E be regarded as falsifying H.

In this way, the cast of mind which will acknowledge only deductive-logical relations between propositions back-fires, so to speak, on its own possessors. It obliges them, on pain of suffering the fate which to them is even worse, of acknowledging non-deductive logical relations, to misuse those very logical expressions which they themselves regard as being the only admissible ones. This is the phenomenon referred to in Chapter Two above, of deductivists being obliged in the end to strangle their own children.

3

THERE are, of course, very few people who believe that deductivism is true. The human race at large is decidedly of the opposite opinion, and holds that there are extremely numerous values of P, Q, and R, such that P is a reason to believe Q, without Q being entailed either by P or by P-and-R for some observational R. Confirmation-theory, or non-deductive logic, or 'inductive' logic as Carnap called it, is the attempt to put into systematic form the very many intuitive beliefs which everyone has about when P is a reason to believe Q. Our authors entertain a boundless hostility and contempt for non-deductive logic; and the explanation of this fact lies, of course, in their deductivism.

Carnap speaks of 'inductive' logic, because he chose to use the word "inductive", as his Glossary indicates,[1] simply as a synonym for "non-deductive". This is a neologism which was apparently unconscious, and which has nothing at all to recommend it. There is nothing to be said for calling the argument, for example, to "Socrates is a man", from "All men are mortal and Socrates is mortal" "inductive"; nor for calling so the argument to "Socrates is mortal" from "99 per cent of men are mortal and Socrates is a man". But there is a great deal to be said against it. It suggests, what is false, that non-deductive logic is concerned exclusively with arguments from the observed to the unobserved; whereas this class of arguments, for all its special importance for empiricist philosophers of science, is only one among many classes of non-deductive arguments. It suggests, what is false, that the thesis of the incurable invalidity of inductive arguments is an analytic triviality; which it is so far from being that Mill and many others, as we saw in Chapter Four, have by implication denied it. And it therefore further suggests, what is also false, that deductivism, and scepticism about induction, are logically equivalent theses; whereas they are, as we have seen, actually independent. These consequences suffice to show that Carnap's neologism was tragically inept. But it has acquired too much currency to be soon reversed, and accordingly I adopt it here; though never without a protest in the form of quotation-marks around "inductive", whenever I use it as an adjective to "logic" or "logicians".

The chief land-marks of 'inductive' logic are Carnap's *Logical Foundations of Probability* (1950), and the articles of Hempel which are collected in *Aspects of Scientific Explanation*.[2] Now these writings, despite both their self-imposed limitations and their consequent essentially fragmentary nature, and despite some positive errors which they undoubtedly contain, represent far more progress, in an area of the first intellectual importance, than the entire history of the human race can show before. Their only serious fore-runners, indeed, are some of the writings which belong to the 'classical' period of the theory of probability, between 1650 and 1850. And what immense strides Carnap, in particular, made, in clarifying, in improving, and in extending, that priceless but profoundly confused historical deposit, many students of probability know; even if others do not. Not contempt, then, but rather all honour, is due to these writers, for the mighty fragments of non-deductive logic which they have left us.

But even if Carnap, Hempel, and their followers, had achieved, as their deductivist critics allege that they have achieved, nothing constructive at all in the way of systematic non-deductive logic, they would still merit the respect which is due to their having been *in earnest* with empiricist philosophy of science. There can be no serious philosophy, of science or of anything else, without seriousness about the logical relations between propositions. And there can be no serious *empiricist* philosophy of *science,* in particular, without seriousness about the *non-deductive* logical relations between propositions. Arguments from the observed to the unobserved really are incurably invalid: *this* much of Hume's philosophy of science is true, and in this much all empiricists are now agreed. But, this much being agreed, any empiricist who is also a *deductivist,* as all our authors are, condemns himself, not just to irrationalism, but to unseriousness, about science.

4

HOSTILITY to non-deductive logic, and the sabotaging of non-deductive-logical expressions, are among the inevitable consequences of our authors' deductivism. But there are also two marked and characteristic features of our authors' writings which, although not inevitable consequences of their deductivism, can only be explained by reference to it, or to that cast of mind which acknowledges only deductive-logical relations between propositions.

One of these features, and one which is at first sight surprising in deductivists, is this: an extreme lack of rigour in matters of deductive logic.

As evidence of this fact, I could of course cite again all the cases, already mentioned in Chapter Two, in which a deductive-logical expression is sabotaged by our authors; all the cases, for example, in which they say that one proposition or kind of proposition 'falsifies' another, when they know well enough that the two are not inconsistent. But obviously it would be preferable, if it is possible, to draw here on entirely independent evidence; and this is not only possible but easy. I will give three instances of the extreme lack of rigour of which I speak. All three are drawn from Popper, who is the most rigorous of our authors.

(i)   Two scientific theories can be inconsistent with one another. This fact is too obvious to need examples to prove it. Nor has any philosopher assumed this obvious truth more often than Popper does. His writings are full of references to incompatible, or conflicting, or competing, or rival, scientific theories. Nor is this an accident. For his entire philosophy of science in fact arose (as we saw in Chapter Three) from contemplating, over and over again as in a nightmare, the overthrow of Newtonian physics by a rival theory; this *kind* of episode in the history of science has always remained his principal concern; and he thinks (as we saw in Chapter Two) that the overthrow of one scientific theory always requires the presence of an incompatible theory.

At the same time, it is an immediate and obvious consequence of the account which Popper gives of the logical form of scientific theories, that one scientific theory *cannot* be inconsistent with another.

This account was given by Popper in his [1959], and has

since then been taken for granted in all his writings. According to it, any scientific theory, and equally any law-statement (for Popper always lumps these two together) is what we may call "a mere denial of existence". That is, it is a proposition which denies the existence of a certain kind of thing, and which does not assert the existence of anything.

"$(x)(Raven\ x \supset Black\ x)$" will suffice as an example of this class of propositions. Since it is logically equivalent to "There are no non-black ravens", it denies the existence of a certain kind of thing; and since, for the same reason, it would be true (as philosophers say) 'in the empty universe', that is, if nothing at all existed, it does not assert the existence of anything. Hence it is a mere denial of existence.

That Popper does conceive scientific theories and laws as mere denials of existence, the following quotation is sufficient to establish. "The theories of natural science, and especially what we call natural laws, have the form of strictly universal statements; thus they can be expressed in the form of negations of strictly existential statements or, as we may say, in the form of *non-existence statements* (or 'there-is-not' statements). For example, the law of the conservation of energy can be expressed in the form: 'There is no perpetual motion machine', or the hypothesis of the electrical elementary charge in the form: 'There is no electrical charge other than a multiple of the electrical elementary charge'. In this formulation we see that natural laws might be compared to 'prescriptions' or 'prohibitions'. They do not assert that something exists or is the case; they deny it. They insist on the nonexistence of certain things or states of affairs, proscribing or prohibiting, as it were, these things or states of affairs: they rule them out."[3]

But now, two mere denials of existence cannot be inconsistent with one another. For in the logically possible case of the empty universe, all such propositions would be true.

Hence Popper, while he constantly assumes that two scientific theories, or two law–statements, can be inconsistent with one another, gives an account of the logical form of such propositions which immediately has, by the most elementary deductive logic, the consequence that they cannot.

(ii)   Let us call the conjunction of Newton's laws of motion with his inverse–square law of gravitational attraction, "Newtonian physics". And let us consider the question whether Newtonian physics in this sense is falsifiable; the question, that is, whether there is any observation-statement which is inconsistent with Newtonian physics.

Even allowing for the differences of detail which exist among philosophers as to what counts as an observation-statement, it is obvious enough that the answer to this question is "no". There can be no *observably* non-Newtonian behaviour, on the part of billiard balls or of anything else. (There could, of course, be non-Newtonian behaviour, for example a billiard ball coming to rest with no forces acting on it; but then, that there are no forces acting on this ball, is a theoretical generalisation, and hence cannot be part of an *observation*-statement.) That Newtonian physics is unfalsifiable, is also evident from the fact that, however oddly billiard balls might behave on a given occasion, Newtonian physics could form part of the deductive *explanation* of this behaviour, by being conjoined with other propositions, perhaps about hidden masses, or about the presence of forces other than inertia and gravitation.

Newtonian physics (in our sense) is evidently a scientific theory. This fact, along with its unfalsifiability, is a refutation of Popper's famous thesis that falsifiability is a necessary condition of a theory's being a scientific one. This criticism of Popper was made by Lakatos.[4]

Popper's reply is given in the following paragraph. "Suppose that our astronomical observations were to show, from tomorrow on, that the velocity of the earth (which remains on its present geometrical path) was increasing, either in its daily or in its annual movement, while the other planets in the solar system proceeded as before. Or suppose that Mars started to move in a curve of the fourth power, instead of moving in an ellipse of power 2. Or assume, still more simply, that we construct a gun that fires ballistic missiles which consistently move in a clearly non-Newtonian track ... There are an infinity of possibilities, and the realisation of any of them would simply refute Newton's theory. In fact, almost any statement about a physical body which we may make—say, about the cup of tea before me, that it begins to dance (and say, in addition, without spilling the tea)—would contradict Newtonian theory. This theory would equally be contradicted if the apples from one of my, or Newton's, apple trees were to rise up from the ground (without there being a whirlwind about), and begin to dance around the branches of the apple tree from which they had fallen, or if the moon were to go off at a tangent; and if all of this were to happen, perhaps, without any other very obvious changes in our environment."[5] (I have here substituted the word "track" where, I take it, "tract" is a misprint in the original.)

That Popper's reference to missiles which move in a "clearly non-Newtonian track" was a flagrant begging of the question, I need hardly state. The question, which Lakatos had answered in the negative, was, precisely, whether there *is* any such thing as a "clearly", that is *observably*, non-Newtonian track.

But the principal defect of the paragraph just quoted is much more simple and amazing than this. Consider the proposition: "There was no whirlwind about; the apples which had fallen from my tree rose from the ground and began to dance round the branches of the tree; and this happened, perhaps, without any other very obvious change in the environment." It might be doubted, in view of the remarkable last clause, whether this proposition is, as it needs to be in order to be relevant at all, an *observation-statement*. But we do not need to decide that. For it is not only obvious, it is blazingly obvious, that *this proposition,* whether it is observational or not, *is not inconsistent with Newtonian physics.* The same is true, and equally obviously true, of the proposition: "The cup of tea in front of Popper began to dance, without spilling the tea." The same is also obviously true of every one of the other examples which Popper gives. Yet he brazenly asserts that any of these propositions, and indeed *"almost any statement about a physical body which we may make ... would contradict Newtonian theory"*.

One can scarcely believe one's eyes while reading this paragraph of Popper. What beginning student of deductive logic would not be ashamed to assert such transparent logical falsities as these? (He would never be tempted to do so, however, because they have not the smallest particle of plausibility to recommend them.) What editor would print such palpable untruths, if they came to him from an 'ordinary philosopher'? It is difficult, in fact, to imagine a more brutal contempt for deductive logic than is displayed by this impudent list of so-called falsifiers of Newtonian physics.

If the paragraph quoted above was not mere bluff (as I believe it was), then it testifies to the survival, in an unlikely quarter, of a belief which was very common in the two preceding centuries, and which has only recently been almost entirely extinguished: the belief that Newtonian physics is a guarantee against the occurrence of—just about anything disagreeable. In the mid–18th century Dr. Johnson refused for six months, on a mixture of Anglican and Newtonian grounds, to believe the reports of the Lisbon earthquake. In the mid-19th century it was widely believed that Newtonian physics, as developed by Laplace in particular, guaranteed the stability and permanence of the solar system (at any rate until 'the trumpet shall sound'). That this belief survived to some extent even up to the mid-20th century, is strongly suggested by the irrational hostility with which Immanuel Velikovsky's theories were received in 1950. Now, of course, when theories like his have become respectable, this belief is almost extinct. But even when it was at its height, say among 18th-century Anglicans, I never heard of anyone who believed that Newtonian physics was a logical guarantee of decent behaviour on the part of his *teacup.*

(iii)   My third example also concerns Newtonian physics, in the same sense as before. In this case the question is, what logical relation does Newtonian physics bear to Kepler's laws of planetary motion?

An answer which has often been given to this question, and which has been still more often implied, is that Newtonian physics *entails* Kepler's laws. It is obvious that this is not so. Kepler's laws entail that the planets and the sun exist; but Newtonian physics has no such entailment.

Popper gives a different answer. Newtonian physics, he says, is actually inconsistent with Kepler's laws. It "formally contradicts"[6] them; "from a logical point of view, Newton's theory, strictly speaking, contradicts both Galileo's and Kepler's ...".[7]

This answer to the above question has become an article of faith among irrationalist philosophers of science. Feyerabend[8], Kuhn[9], and many others[10] repeat it. Yet it is obvious that this answer too is false: *strictly speaking,* and *formally,* Newtonian physics is *not* inconsistent with Kepler's laws.

Kepler's laws are purely kinematic; that is, they simply ascribe certain motions to certain bodies, and say nothing whatever about the mass of anything, or about any force exerted by or on anything. And it is easily seen that no purely kinematic proposition is inconsistent with Newtonian physics. Take any purely kinematic proposition: for preference, here, such a highly 'non-Newtonian' one as, say, "The planets describe rectangles with the sun as centre". This proposition is so far from being inconsistent with Newtonian physics, that it could be *deduced from and explained by* Newtonian physics, in conjunction with certain contingent assumptions about the forces to which the planets are subjected, the mass of the planets, their cohesiveness, and so on. Some at least of these auxiliary premises would of course be false in fact. Their conjunction with Newtonian physics would therefore be false in fact too. But that conjunction, obviously, need not be *logically* false; as it would have to be, if the hypothesis of rectangular orbits were actually inconsistent with Newtonian physics. And the same is equally true of any other purely kinematic proposition, such as Kepler's laws.

THE examples which have just been given, of carelessness, or something a good deal worse, regarding deductive-logical relations between propositions, are characteristic of our authors. They are in fact characteristic of them in two senses. One is, that other examples of the same kind can easily be supplied from their writings.[11] The other is, that there is no parallel to these examples in the writings of non-irrationalist philosophers of science. In particular, one would look in vain in the writings of the 'inductive' logicians, for anything corresponding to the carelessness, in matters of deductive-logical relations between propositions of science, which has just been illustrated from our deductivist authors.

Now this is, at least at first sight, surprising. One expects a deductivist to be a severe judge of logical morals, and not only of other people's: his own logical conduct one would expect to be above reproach. He is the last person one would expect to assert or imply that a certain deductive-logical relation exists, in cases where it does not, or does not exist, in cases where it does. And on the other hand such sloppiness would not be at all unexpected from 'inductive' logicians. After all, they are 'soft on logic'. For example they wish to palliate those inductive propensities which erring man shares with the brute creation, and which deductivists think should only be reprehended. Indeed, the entire enterprise of 'inductive' logic appears (at least to its critics) intended to conceal, by a fig-leaf of system, the naked indecency of affirming the consequent. Why then is it precisely the *deductivists,* and *not* the 'inductive' logicians, whose deductive logic turns out to be bad? Why is it that, while Popper's philosophy of science furnishes a steady stream of examples of indifference to elementary deductive logic, the philosophy of Carnap or of Hempel does nothing of the kind? I believe I can answer this question.

Deductivism, it is to be remembered, is a variety of perfectionism: it is an 'only the best will do' thesis. And, at least in very many domains, perfectionism is especially apt to produce performance which is actually further from perfection than the average for that domain. In politics, for example, perfectionism is wisely recognised as having brought into being the very worst societies. In philosophy each of us knows some one whose standards are so extremely high that he never does any philosophy at all. In morals the ancient perfectionist doctrine, often revived, that all evils are *equally* evil, is at once recognised by any person of common sense as sure to have disastrous moral effects in practice. And so on.

Nor is the inner mechanism of these causal connections at all hard to perceive. It is this. The perfectionist, by his exclusive concentration on the ideal, is prevented from attending to the *differences* which exist among cases in which that ideal is not satisfied: even though such cases may include all the actual ones (the ideal being so high), and even though the differences are very great between some of these cases and others.

This gives us a reason to anticipate that deductivists will be, in practice, uncommonly careless in matters of deductive logic. But it does not on its own quite cover the particular case before us. For here the errors are all of one particular kind, and of a kind which is still rather surprising. Our authors' carelessness never takes the form of implying that some specific deductive-logical relation does not exist, in cases where it does. It always takes the form, as in the three examples given above, of implying that a specific deductive-logical relation exists, in cases where it does not. This cannot be explained just by the tendency of perfectionism to result in *general* carelessness. On the contrary, it is somewhat surprising in perfectionists. For its moral analogue (for example) would be, a perfectionist whose neglect of the differences between actual evils often carried him to the point of positively *mistaking* actual evils for ideal goods. This is not what one expects from perfectionists. How is this particular *kind* of carelessness to be explained?

Popper, as we saw in Chapter Two above, rejected the belief that there are propositions of science "the analysis of [whose] relations compels us to introduce a special probabilistic logic which breaks the fetters of classical logic".[12] In opposition to such views, he undertook to show that, even in the apparently intractable cases, the logical relations between propositions of science can always be "fully analysed in terms of the 'classical' logical relations of *deducibility* and *contradiction*".[13]

Well, let us consider what would be the result of *actually* carrying out this undertaking. What is the *actual* 'classical' logical relation between, for example, one scientific theory and another, in those cases in which they are intuitively and rightly called "competing" or "rival" theories? The relation cannot be contradiction or contrariety. For as we have seen, two scientific theories (taking their logical form to be what our authors say it is) cannot be inconsistent. The relation cannot be subcontrariety, since any two rival scientific theories might both be false. There cannot be logical equivalence between them, or over-entailment either way; for in any one of those three cases it would be quite wrong to call the two theories *competing*. But these six are the only deductive-logical relations possible between two contingent propositions, apart from independence. The logical relation, therefore, between any two competing scientific theories, is independence.

What is the classical logical relation between, for example, Newtonian physics and any observation-statement? Not inconsistency, as we have seen, Not subcontrariety, since both could be false. There is no entailment either way. So the answer to this question, too, is: independence.

What is the logical relation between Newtonian physics and Kepler's laws of planetary motion? Not inconsistency, as we have seen. Not subcontrariety, since both could be false. There is no entailment either way. So the 'classical' answer to this question, too, is independence.

What is the classical logical relation (to go back to a class of examples which was, as we know, of peculiar importance to Popper) between H, "The probability of a human birth being male is .9", and E, "The observed relative frequency of males among births in human history so far is .51"? Again the answer is, of course, independence.

Evidently, this is going to be an excessively uninteresting philosophy of science! Yet the four questions just asked are ones intensely interesting to any philosopher of science, and are in fact typical of the questions which interest him. But if he insists on confining his answers to classical or deductive-logical relations, then the only answer which he can give with truth to any of them is the uninteresting one, "independence". This answer is uninteresting, because almost *any* two propositions are logically independent: for example, almost any two non-competing scientific theories (the Copernican theory and Darwinism, say), as well as any two competing theories, are independent. And the ultimate reason for *that,* of course, is that the classical logical relation of independence is extremely *unspecific:* it comprehends indifferently logical relations which are in fact of the utmost diversity. Hence by giving just this same answer in all of the four examples above, obvious and important logical differences among those cases are suppressed: the fact, for example, that while in the last case given above, E *disconfirms* H, Kepler's laws do not disconfirm Newtonian physics.

Confronted, then, with almost any interesting question about the logical relations between propositions of science, a philosopher who is resolved to confine his possible answers to deductive-logical relations, is faced with an extremely painful choice. He absolutely must: either give no answer at all; or give an answer which is true but is excessively uninteresting both to himself and others; or give an answer which may be interesting but is false.

Now our authors, as I have said, always in fact choose the third of these alternatives. This is explicable, but on only one hypothesis, to the nature of which the first two alternatives just mentioned provide the clue. For the three alternatives may be reduced just to the following two: a deductivist philosopher must either give a false answer, or *suffer painful under-exercise of his logical faculty.* Our authors' characteristic kind of carelessness, of attributing to a pair of propositions a deductive-logical relation which they simply do not possess, is therefore a case of *vacuum-activity* in Lorenz's sense.[14]

The commonest case of vacuum-activity is that in which a dog, long deprived both of bones and of soil, 'buries' a non-existent bone in non-existent soil (usually in the corner of a room). This behaviour-pattern is innate in dogs, and if deprived for too long of its proper objects, it simply 'discharges' itself in the absence of those objects. After a certain point, bone-free life is just too boring for dogs.

Just so, our authors are philosophers of science, and have a built-in need to answer interesting questions about the logical relations between propositions of science. But what can be said with truth in answer to such questions, *without 'breaking the fetters of classical logic'*, is painfully uninteresting; while our authors are resolved to permit themselves no other kind of answer. After a certain point, however, life without interesting logical relations is just too boring for philosophers. Sooner or later, then, *another* and more interesting deductive-logical answer discharges itself, although in entire disregard of the absence of its proper objects. Then, for example, Kepler's laws and Newtonian physics are called "inconsistent"; although anyone who is not under the same compulsion as the deductivist easily sees at once that those two propositions are in fact merely independent.

This phenomenon can equally well be looked at, of course, from the other end. The dog engaged in his vacuum-activity, if he could write, might say, exactly in the style of Lakatos, "I am 'burying' a 'bone'." If he were more Popperian he might write either "I introduce a methodological rule permitting us to regard this as bone-burying", or "Any similarly-deprived dog would regard this as burying a bone". His ghost-behaviour corresponds to their ghost-logical statements.

The apparent paradox, of deductivists whose deductive logic is sloppy, and 'inductive' logicians whose deductive logic is not, is thus resolved. Our authors, by their determination to acknowledge no other than deductive-logical relations, are self-condemned, when they come to almost any interesting question about the relation between propositions of science, to being totally silent, totally uninteresting, or totally wrong. Faced with the first two dread alternatives, a philosopher's reaction will not be long in doubt. The 'inductive' logician, on the other hand, is from the start under no such compulsion. That is why he can write about the relations between propositions of science, without having to produce a stream of elementary mistakes in deductive logic.

5

THERE is another feature of our authors' writings, a feature even more pronounced and more characteristic of them than their carelessness about deductive logic, the explanation of which also lies in their deductivism. This is, their levity or *enfant-terriblisme*.

The levity of Feyerabend is too 'gross, open, palpable', to require that instances be given here to prove it. In *Against Method* it is in fact so omnipresent that he has managed to entangle himself in a certain 'paradox of levity' which is, as far as I know, entirely original. Feyerabend enjoins the reader of that book[15], indeed he pleads with him[16] , not to take anything he reads there too seriously. But this injunction and this plea are among the things he reads there. How seriously, then, ought the reader to take *them*?

Lakatos's fame as a philosopher of science rests principally on his [1970]. He there claimed, among other things, to give an account, more accurate than anyone else had given, of the actual *history* of science. Yet that essay contains several episodes from the history of science, episodes complete with circumstantial detail, which are, Lakatos calmly tells us in footnotes[17], fabrications of his own. Perhaps it will be said that this instance is not characteristic: a mere isolated outcrop of levity. Even if it were so, this particular way of wasting paper is not one which would even suggest itself to a philosopher who was in earnest with his subject. But in any case there are in Lakatos many instances of levity which are indisputably characteristic. One is the long footnote in *Proofs and Refutations* about proof in mathematics. This begins with the remarkable understatement, that "Many working mathematicians are puzzled about what proofs are for if they do not prove." And Lakatos goes on to quote, with relish, the mathematician G.H. Hardy, as follows: " 'proofs are what … I call *gas,* rhetorical flourishes designed to affect psychology, pictures on the board in the lecture, devices to stimulate the imagination of pupils.'"[18] As academic humour, this may be allowed to pass (combining as it does self-contempt, and the contempt of others, in the prescribed unequal proportions); as serious philosophy of mathematics, not.

Many people suppose that Popper is far removed, at least in this matter of levity or *enfant-terriblisme,* from those intellectual progeny of his to whom I have just referred. Indeed, there are at the present time many youngish philosophers of science in whose writings Popper is made to serve (since they know next to nothing of any philosopher of science before him) as their standing and cautionary example of the unbearable *gravitas* which characterised philosophers of science in the dark ages. But this is merely a measure of the ignorance of such persons; the truth is exactly the opposite. *The Logic of Scientific Discovery* was no less an *enfant-terrible* first book than *Language, Truth and Logic,* or *A Treatise of Human Nature.*

The simple and sufficient proof of Popper's levity is this:

that he is always saying 'daring' things that he does not mean. For example he says, and says, as we have seen, with all possible emphasis, that there is no good reason to believe any scientific theory. But he is not in earnest. He does not really believe that there is no good reason to believe that his blood circulates, or that the earth rotates and revolves, or that his desk is an assemblage of molecules—or a thousand other scientific theories which could as easily be mentioned. Confront him with members of the Stationary Blood Society, who *are* in earnest when they say there is no good reason to believe that the blood circulates, and Popper would find the difference manifest enough between real irrationalism, and his own 'parlour-pink' version of it. Indeed, even as things are, Popper every now and then notices, to his alarm, that what Hume called 'the rabble without doors' shows some tendency to *agree* with him, that there is no good reason to believe any scientific theory: at these points, the reader of Popper is about to receive another lay-sermon on the deplorable growth of irrationalism, relativism, etc. In other words, Popper's daring irrationalist sallies are meant to be tried, like a baron under Magna Charta, only by a jury of his peers, and for the same reason: that other people might not *understand*.

The levity of Popper and his followers concerning science bears a marked analogy, therefore, to a species of *political* levity which is excessively familiar: what Kipling called "making mock of uniforms that guard you while you sleep". For who are the pet aversion of Popperites, as policemen are of parlour-pinks? Why, ordinary flesh-and-blood scientists, of course! Any contact with living scientists always leaves a Popperite far more Feyerabendian than it found him. It can be relied on to bring him out in a fury of what we may call 'criticismism'. Scientists, he finds to his horror, are dogmatic, uncritical, authoritarian, etc., etc. So they are, of course. They are also people of the very same kind, by and large, as those who have erected what Popper himself once called, in a moment of self-forgetfulness, "the soaring edifice of science".[19]

It is the frivolous elevation of 'the critical attitude' into a categorical imperative of intellectual life, which has been at once the most influential and the most mischievous aspect of Popper's philosophy of science. That it is frivolous, should be evident from the tautology that it is only valuable criticism which is of value; not criticism as such. The demand that scientists *in general* should be critics and innovators, rather than mere followers, is even, in its extreme forms, selfcontradictory; like the implicit demand of those educationalists who want every child to be *exceptionally* creative. (Before they complain of the rarity of any great critical faculty in scientists, Popperites should read Hume on what he called those "thoughtless people" who complain of the rarity of great beauty in women.[20]) Even in its non-extreme forms, however, the apotheosis of the critical attitude has had, as its principal effect, simply this: to fortify millions of ignorant graduates and undergraduates in the belief, to which they are already only too firmly wedded by other causes, that the adversary posture is all, and that intellectual life consists in "directionless quibble".[21]

IN the most marked contrast possible to all of this, the writings of the 'inductive' logicians are entirely free from levity. These philosophers take science, not as furnishing the materials for mere 'critical discussion', but seriously. They have nothing of our authors' bohemian contempt for, or disbelief in, success. On the contrary, scientific success is treated by them as the obvious though wonderful fact which it is. They never say daring things that they do not mean about science.

The levity of our authors, and the absence of levity in the 'inductive' logicians, is sufficiently obvious as a fact. But why do I say that the explanation of it lies in the fact that our authors are deductivists, while the 'inductive' logicians are not? The main part of my answer is this: that deductivism, the thesis from which all the disagreements between these two groups of philosophers spring, is a proposition which can recommend itself *only* to the minds of *enfants-terribles* or other extreme doctrinaires, and more specifically, that deductivism is a thesis of an intrinsically *frivolous* kind.

Consider the argument to the conclusion "I will win a lottery tomorrow", from: "There is a fair lottery of 1000 tickets, to be drawn tomorrow, in which I hold just one ticket or none". Here everyone would agree that the premise is no reason to believe the conclusion. Anyone who said the same thing, however, about the argument to the same conclusion from the above premise minus its last two words, would find few to agree with him. On the contrary, the difference in logical value between the two arguments is so manifest, that such a person would be thought to display an almost unheard-of degree of logical blindness or perversity. But let us change the premise again, so that it now ends with "... in which I hold just 999 tickets". Anyone who said that, even here, the premise is no reason to believe the conclusion, would evidently thereby announce himself as one of those hopeless doctrinaires with whom rational argument, and even 'critical discussion', is effort thrown away.

The deductivist, however, must say that in all three of these arguments the premise is no reason to believe the conclusion. For all three are invalid, and incurably so. This is enough to show that deductivism is one of those theses which, although anyone under pressure of philosophical argument might momentarily reconcile himself to it, would not be adhered to willingly and with knowledge of its consequences, by anyone except an *enfant-terrible* or an extreme doctrinaire.

Suppose I have come to know that P, "I hold just 999 of the 1000 tickets in a fair lottery to be drawn tomorrow"; and suppose that, as a result of acquiring this knowledge, I have come to have a higher degree of belief than I had before in the proposition Q: "I will win a lottery tomorrow". Suppose that I am then reminded by some one of the fact that R, "It is logically possible that P be true and Q false"; and suppose I fully accept this truth, and add it to my stock of knowledge. I acknowledge, in other words, that although I hold nearly all the tickets in this fair lottery, I might not win it. Suppose, finally, that on account of adding this truth R to my premise P, I com e to have a *lower* degree of belief in Q than I had before being reminded of R.

In that case, it will be evident, I am being irrational, and more specifically I am being frivolous. Irrational, because R is a *necessary* truth, and hence its conjunction with P is logically equivalent to P itself, while two arguments cannot differ in logical value if their premises are logically equivalent and they have the same conclusion. And my irrationality is of a frivolous kind. My conclusion Q is a contingent proposition, saying only that the actual world is thus-and-so. My additional premise R is a proposition true in all possible worlds. But a proposition true in all possible worlds cannot tell in the slightest degree for or against any proposition just about the actual world. (If it could, why ever leave the armchair at all? Why not do *all* our science *a priori?*) Yet after having allowed my degree of belief in the contingent Q to be raised by the contingent P, I have allowed it to be depressed again by the addition to P of a premise R which, where the conclusion of the argument is contingent, as it is here, *cannot weigh anything at all*. To do this is *light-mindedness* on my part; and it would be light-mindedness in anyone else to demand it of me.

Let us change the example to one in which the argument

is inductive. P is now "All the many flames observed in the past have been hot", and Q is "Any flames observed tomorrow will be hot". Suppose that I have come to know P, and that, as a result of acquiring this knowledge, I have come to have a higher degree of belief in Q than I had before. Suppose I am then reminded by some one of the fact that R, "It is logically possible, however many may be the 'many flames' referred to in P, that P be true and Q false". And suppose that I fully accept this truth, and add it to my stock of knowledge.

Now, if on account of adding this truth R to my premise P, I come to have a lower degree of belief in Q than I had before, then I am being irrational in exactly the same frivolous way as in the case of the lottery. For here too the additional premise R is a necessary truth, while the conclusion of the argument Q is contingent. Therefore R cannot tell in the slightest degree against or for Q. Yet having allowed my degree of belief in the contingent Q to be raised by the contingent P, I have allowed it to be depressed again by the addition to P of a premise R which cannot weigh anything at all in an argument about whether flames will be hot tomorrow.

Yet it is *precisely* this piece of light-mindedness that the deductivist demands of me. The deductivist, Hume for example, tells me that P is no reason to believe Q; and of course, if that *is* so, then I should indeed lower my degree of belief in Q. But, I ask him, *why* is P no reason to believe Q, or *why* should I lower my degree of belief in Q? Is Hume about to remind me of some quite other contingent fact S, which I have neglected, and which tells *against* Q, perhaps even making it probable that some flames tomorrow will *not* be hot? Hardly! So I repeat my question: why should I lower my degree of belief in Q? Forsooth, Hume tells me, just for this reason: that a man who infers Q from P, or from P conjoined with any other observation statement, "is not guilty of a tautology";[22] that given P, and any other observational premise, "the consequence [Q] seems *nowise necessary*";[23] that, whatever our experience has been, "a change in the course of nature ... is *not absolutely impossible*";[24] that past and future hot flames are 'distinct existences', that is, that the one *might* exist without the other; and so on.

This, and nothing else in the world, is what Hume finds to object to in my inductive inference from P to Q. This is the whole of his answer to the question, why I should lower my degree of belief in Q. Yet it amounts just to this, that the inference from P to Q is invalid, and remains so under all observational additions to its premises; or in other words just to R, that it is possible for P, and any other observation-statement, to be true, and Q false. But this is a necessary truth. And therefore to demand, just on *this* account, that I should lower my degree of belief in the hotness of tomorrow's flames, is mere frivolity.

Of course exactly the same is true of Popper. If I have, as Popper says I should not have, a positive degree of belief in some scientific theory, what can Popper urge against me? Why, nothing at all, in the end, except this: that despite all the actual or possible empirical evidence in its favour, the theory *might* be false. But this is nothing but a harmless necessary truth; and to take it as a reason for not believing scientific theories is simply a frivolous species of irrationality. Yet it is this proposition, that any scientific theory, despite all the possible evidence for it, *might* be false: a proposition loudly announced by the fall of Newtonian physics; amplified ever since by morbidly sensitive philosophic ears; endlessly re-applied and re-worded; insisted on to the exclusion of every other logical truth about science, and mistaken for a reason for not believing scientific theories: it is this proposition, so treated, which may be said to *be* recent irrationalist philosophy of science.

This phenomenon is so far from being new, that it appears to be a perennial feature of sceptical or irrationalist philosophy. To furnish a reason for doubting all contingent propositions among others, Descartes appears to have thought it sufficient if he could establish the *logical possibility* of an all-deceiving demon.[25] The sceptics of later classical antiquity were fully conscious of the dependence of their entire philosophy on expressions such as "might" and "possibly", and they appear to be constantly guilty of taking logical truths involving such expressions as grounds for doubting contingent propositions.[26] And among recent irrationalist philosophers of science, along with neutralised success-words and sabotaged logical expressions, an unfailing literary diagnostic is, the use of the frivolous or deductivist "might". Such philosophers can be absolutely relied on to try to cast doubt on the truth of contingent propositions, by the enunciation of mere logical truths about the *possibility* of their falsity.

In Hume's *Treatise, Abstract,* and first *Enquiry,* deductivism, conjoined with the incurable fallibility of induction, led to scepticism about induction. The latter two books were, of course, re-writings of Book I of the *Treatise:* "a juvenile work", as Hume tells us, "which the Author had projected before he left College".[27] It is therefore not surprising that in the central argument of those books, concerning induction, the key premise should have been the *enfant-terrible* thesis of deductivism. But Hume, unlike our authors, did not remain a deductivist *enfant-terrible all* his life. In the one philosophical work of his maturity, which is also his best, the *Dialogues concerning Natural Religion,* the incurable invalidity of induction is maintained as firmly as ever. But at the same time, in that book, inductive *scepticism,* and therefore by implication deductivism, are rejected very early, and with a summariness which is well-proportioned to their frivolity.[28]

Hume did better than that, however. Late in his life he made precisely the contemptuous dismissal that any rational inductive fallibilist must make of inductive scepticism, and by implication of deductivism. This was on his deathbed, in a conversation with Boswell on the subject of immortality. Boswell, almost desperate for some hint of consolation, "asked him if it was not possible that there might be a future state. He answered, It was *possible* that a piece of coal put on the fire would not burn; and he added that it was a *most unreasonable fancy* that he should exist for ever."[29] I earnestly commend this remark, of their founding father and favourite deductivist, to irrationalist philosophers of science. For it, and not the deductivist levity of the *Treatise* or of their own writings, expresses exactly the response of a rational man to contingencies which are recommended to his belief just on the impertinent ground of their possibility.

Of course I do not say that every philosopher who is a deductivist is frivolous. I do say that deductivism is intrinsically a thesis of a deeply frivolous nature; that it is the premise from which flow all the irrationalist consequences of our authors' philosophy of science; and that the levity which their philosophy exhibits so markedly is therefore to be explained, as their irrationalism is, by the influence on their minds of deductivism. But it is no more to be inferred from the fact that deductivism is frivolous, that all deductivists are frivolous, than it is to be inferred from the fact that patience is a virtue, that all the patient are virtuous. And in the one case or in the other, the conclusion would be false in fact; for even among our four authors there is one who, in this respect, stands apart from the others.

Kuhn shares with our other authors, as he must, their boundless contempt for 'inductive' logic. His remark about "cloud-cuckoo land", for example, quoted in Chapter Two above (see the text to footnote 31), is a thinly-veiled contemptuous reference to it. But setting this point aside, his writings are entirely free from the levity which disfigures the writings of our other authors. His philosophy of science is not daring; only shocking. He has no time at all for criticismism, and to *épater les bourgeois* is the least of his concerns. His admiration for 'normal science' is so pronounced that it brings out Popper and his followers in a perfect rash of Spocks.[30]

The reason is, that Kuhn is in earnest with irrationalist philosophy of science, while the others are not. He actually believes, what the others only imply and pretend to believe, that there has been no accumulation of knowledge in the last four centuries.[31] And he even bids fair, by the immense influence of his writings on 'the rabble without doors', to make irrationalism the *majority* opinion. 'This was the most unkindest cut of all' for our other authors, and is in fact the real ground of the offence which Kuhn has undoubtedly given them; as distinct from the avowed but manifestly spurious ground mentioned at the beginning of this book. For the cruellest fate which can overtake *enfants-terribles* is to awake and find that their avowed opinions have swept the suburbs.

THERE are, unfortunately, grounds for believing that the deductivist cast of mind is, like priesthood, indelible; or at least that deductivism, and the levity which is its natural consequence, can never be entirely erased from any mind in which they have once taken hold.

Consider again the fair lottery to be drawn to-morrow, in which I hold just 999 of the 1000 tickets. Imagine this case to be described by a contemporary philosopher: one who was formerly a deductivist, but who has since 'put away the toy trumpet of sedition' in philosophy. This philosopher, in other words, has arrived at that prodigious pitch of learning which enables him to say, and to believe, what non-philosophers believed all along: that, in this case, while it is possible that I will not win a lottery to-morrow, it is probable that I will.

Now, will there not be, even so, a faint apologetic smile accompanying the word "probable", but not the word "possible", if our ex-deductivist is speaking? If he is writing, will there not be sabotaging quotation-marks around "probable", though not around "possible"? Almost to a certainty there will. A contemporary philosopher can hardly rid himself, even if his life depended on it, of the feeling that the possibility of my not winning the lottery is 'objective', in some sense in which the probability of my winning it is not.

It is essentially the same in the inductive case. The contemporary philosopher will admit easily enough, once it is pointed out to him, that it is a mere *logical* truth that tomorrow's flames *may* be unlike past ones; and that therefore this *cannot* be a reason to *doubt* that they will be like them. Yet in spite of all his efforts to prevent it doing so, this logical truth operates on his mind as though it *were* such a reason, and a weighty one. So obsessive is our endless re-enactment of the death of Newtonian physics, and so permanently disabling is 'modern nervousness' in the philosophy of science.

That this state of mind is a confused one, it can hardly be necessary to say. Probabilities are no less objective than possibilities. On any philosophy of probability, alternatives which are equally probable can be called, with equal propriety, equally possible; and for one alternative to be more probable than another, it is logically sufficient that there be, for every way in which the second can be realised, an equally possible way in which the first can be realised, but not conversely. Anyone, therefore, who is hypersensitive to possibilities, but at the same time is insensitive to all differences of magnitude between probabilities, is certainly in a deeply-confused mental state; even, one would think, a pathological state. Yet this is, in some degree, the actual mental state of most philosophers of science at the present time, and is to a pre-eminent degree the mental state of deductivist philosophers of science, such as our authors.

If it is true that any philosopher who was once a deductivist will carry at least some tincture of deductivism to his grave, then the prospects are so much the worse for there being any future philosophy of science which is free from the levity and other vices of irrationalism. For there can scarcely be any contemporary philosopher of science who is not either a deductivist or an ex-deductivist.

AFTERWORD

A s might have been predicted, the reaction in the philosophical world to the original edition of this book, *Popper and After: Four Modern Irrationalists,* Pergamon Press, 1982, was polarised. David Papineau, a leading British philosopher of science, wrote in the *Times Literary Supplement*: 'Stove has got Sir Karl Popper exactly right ... *Popper and After* will serve as an excellent antidote for the many philosophical innocents who are still in danger of being bewitched by Popper.' Another leading philosopher of science, Joseph Agassi, returned the book to the *British Journal of Sociology of Education* as 'unreviewable', but his letter of complaint was printed in place of a review. An enraged reviewer for *New Scientist*, 19 January 1984, contacted Popper to ask whether he believed in the advance of scientific knowledge. Sir Karl of course replied in the affirmative, saying that the question was like suggesting to a Rothschild that he didn't believe in the existence of money. No hard questions were ventured, however, as to the compatibility of that position with Popper's other expressed views.

Philosophy of Science, the leading journal in the field, gave a generally positive review, as did *Philosophical Books*. A number of other reviewers expressed distaste for the polemics, but conceded that Stove had identified deductivism as a major difficulty for Popper and the others. Typical was the *British Journal for the Philosophy of Science*, which found it 'impossible to recommend this unfortunate book', but agreed that Popper had found no way of reconciling deductivism with the advance of science.

The 'four modern irrationalists' did not deign to reply in print, but a substantial consideration of the book from a Popperite point of view can be found in J. Watkins, 'On Stove's book, by a fifth "irrationalist" ', *Australasian Journal of Philosophy*, 63 (1985): 259-68, and more briefly in D. Miller, *Critical Rationalism*, Open Court, Chicago, 1994, pp. 52-4. Other discussions are: S. Yates, 'Stove's critique of "irrationalists" ', *Metaphilosophy*, 18 (1987): 149-160; R. Sylvan, 'Science and science: relocating Stove and the modern irrationalist', *Research Series in Unfashionable Philosophy*, 1 (1984): 35-54; S.C. Hetherington, 'Stove's new irrationalism', *Australasian Journal of Philosophy*, 76 (1998): 244-9. Much of this debate appeared in Polish in K. Jodkowski, (ed.), 'Na czym polega racjonalnosc nauki', *Realizm, Racjonalnosc, Relatywizm*, vol. 7, MCS Publishing House, Lublin, 1991. The book is summarised in J. Srzednicki and D. Wood, eds., *Essays on Philosophy in Australia*, Kluwer, Dordrecht, 1992, pp. 242-5, 282-4. A complete list of reviews is in *Cricket versus Republicanism*, p. 142.

Some support from scientists came in T. Theocharis and M. Psimopolous, 'Where science has gone wrong', *Nature*, 329 (1987): 595-8, and A.J.M. Garrett, 'Probability, philosophy and science', in J. Skilling, (ed.), *Maximum Entropy and Bayesian Methods*, Kluwer, Dordrecht, 1989, pp. 107-16.

Further reaction was provoked by Stove's article, 'Karl Popper and the Jazz Age', *Encounter*, 65 (1) June 1985: 65-74, (reprinted in *The Plato Cult*). It summarised the argument of *Popper and After*, and suggested that Popper's irrationalist philosophy of science was a product of the reversal of values characteristic of the Jazz Age, along with Freudianism, Cubism, Dadaism and Jazz itself. Stove writes: 'Cole Porter's words

"Anything goes" are not quite right for this situation, though; for they suggest *random* change, or anarchy. He is nearer the mark with "day's night to-day", "good's bad today", and so on; for these words convey the idea of *reversal* rather than of random change.' The article generated a large correspondence, and a not-for-publication letter from Popper, along the lines of 'sad that a prestigious journal such as yours ... '

The inability of the scientific world to respond to Feyerabend's attack was compared to the West's powerlessness in the face of Communist subversion in Stove's 'Paralytic epistemology, or the soundless scream', *New Ideas in Psychology*, 2 (1984): 21-24, (reprinted in *Cricket versus Republicanism*). He published a more technical book on the justification of inductive reasoning by probability theory in *The Rationality of Induction*, Clarendon, Oxford, 1986.

Stove's ideas on the objective nature of probabilistic support for theories were applied to mathematical conjectures in J. Franklin, 'Non-deductive logic in mathematics', *British Journal for the Philosophy of Science*, 38 (1987): 1-18, and to the evidence for historical theories in K. Windschuttle, *The Killing of History*, Free Press, New York, 1997, chapter seven. P. Forrest's *The Dynamics of Belief*, Blackwell, Oxford, 1986, developed them as part of a general theory of how systems of belief adapt to new evidence.

Views of the nature of science similar to Stove's have remained rare in the philosophy of science. The school of 'subjective Bayesians' represented by J. Earman's *Bayes or Bust?*, MIT Press, Cambridge, Mass, 1992, and C. Howson and P. Urbach's *Scientific Reasoning: The Bayesian Approach*, Open Court, Chicago, 1993, has, it is true, kept alive the idea of probability as a foundation of scientific inference. But they have defended the objectivity only of the *principles* of probability, and have maintained that the basic degree of belief of a particular theory on a particular body of evidence may be chosen by the user. Closer to Stove's point of view, though more involved in technical questions, is the school of 'objective Bayesians', led until his death in 1998 by E.T.Jaynes.The publication of Jaynes' book, *Probability Theory: The Logic of Science*, is eagerly awaited.

Such ideas have found a more receptive audience among workers in Artificial Intelligence, who are faced with the problem of actually implementing some workable scheme of inference under uncertainty. Introductory material may be found in the articles on 'Reasoning, plausible', 'Reasoning, nonmonotonic' and 'Probabilistic networks' in *The Encyclopedia of Artificial Intelligence*, ed. S.C. Shapiro, 2nd ed., Wiley, New York, 1992.

There are three later books by Stove on other topics. *The Plato Cult and Other Philosophical Follies*, Blackwell, Oxford, 1991, is an attack on idealism and a range of other diseases of philosophy. Its chapter on philosophers who believe it possible literally to 'make worlds with words' anticipates much later criticism of postmodernism as a revival of philosophical idealism. *Darwinian Fairytales*, Avebury Press, Aldershot, 1995, attacks both Darwin's and recent sociobiologists' versions of evolutionary theory. Stove's polemical essays on such topics as feminism, race and religion were collected in *Cricket versus Republicanism and Other Essays*, Quakers Hill Press, Sydney, 1995. It includes an appreciation and brief biography of Stove, and a bibliography of his work.

JAMES FRANKLIN
AUGUST 1998

NOTES

To Foreword

1. Diana Dyason, 'After Thirty Years: History and Philosophy of Science in Australia 1946-76', *Melbourne Studies in Education 1977*, Melbourne University Press, Melbourne, 1977; Larry Laudan, 'Thoughts on HPS: 20 Years Later', *Studies in History and Philosophy of Science*, 20, 1, 1989, pp 9-13.
2. C. P. Snow, *The Two Cultures and the Scientific Revolution*, Cambridge University Press, Cambridge, 1959.
3. Thomas Kuhn, *The Structure of Scientific Revolutions*, University of Chicago Press, Chicago, 2nd edn., 1970.
4. E. Garfield, 'A different sort of great books list: the 50 Twentieth-century works most cited in the *Arts and Humanities Citation Index*, 1976-1983,' *Arts and Humanities Citation Index*, second semiannual, 1994, pp 7-11; also *Philosophy*, 70, 1995, p 609.
5. Thomas Kuhn, *The Structure of Scientific Revolutions*, p 150.
6. Thomas Kuhn, *The Structure of Scientific Revolutions*, pp 94, 153.
7. David Bloor, *Knowledge and Social Imagery*, Routledge and Kegan Paul, London, 1976.
8. Bruno Latour and Steve Woolgar, *Laboratory Life: The Social Construction of Scientific Facts*, Sage, London, 1979, pp 17, 284.
9. H. M. Collins, *Changing Order: Replication and Induction in Scientific Practice*, Sage, London, 1985.
10. Paul Feyerabend, *Against Method: Outline of an Anarchistic Theory of Knowledge*, New Left Books, London, 1975.

11. Paul Feyerabend, *Against Method,* p 299.

12. Paul Feyerabend, *Against Method,* p 308.

13. 'The Worst Enemy of Science', *Scientific American,* May 1993, pp 16-17.

14. Thomas Kuhn, *The Structure of Scientific Revolutions,* p 149.

15. R. W. Connell, 'Notes on the World Intelligentsia', *The UTS Review: Cultural Studies and New Writing,* 3, 1, May 1997, p 77. The reference to Islamic science that Connell cites is: Ziauddin Sardar, *The Touch of Midas,* Manchester University Press, Manchester, 1984.

16. Mark Walker, *Nazi Science: Myth, Truth and the German Atomic Bomb,* Plenum Press, New York, 1995.

17. David Joravsky, *The Lysenko Affair,* Harvard University Press, Cambridge MA, 1970; Dominique Lecourt, *Proletarian Science? The Case of Lysenko,* New Left Books, London, 1977; Jasper Becker, *Hungry Ghosts: China's Secret Famine,* John Murray, London, 1996. Mao Tse-Tung used Russian theory to draw up eight rules for Chinese farming, one of which commanded that seedlings should be planted much closer together than before. The theory that they would die if crowded was relegated to the competitive assumptions of bourgeois science whereas the Great Helmsman's proletarian theory claimed plants of the same 'background' would fraternally share light and food.

18. Karl Popper, *The Logic of Scientific Discovery,* Hutchinson, London, 1959.

19. Karl Popper, *Conjectures and Refutations: The Growth of Scientific Knowledge,* Routledge, London, 1963.

20. Norwood Russell Hanson, *Patterns of Discovery,* Cambridge University Press, 1958; A. F. Chalmers, *What Is This Thing Called Science?,* (1976), 2nd edn., University of Queensland Press, 1982, pp 22-37.

21. Thomas Kuhn, *The Structure of Scientific Revolutions,* p 146-7.

22. Imre Lakatos, 'Falsification and the Methodology of Scientific Research Programs', in I. Lakatos and A. Musgrave (eds.), *Criticism and the Growth of Knowledge,* Cambridge University Press, Cambridge, 1970.

23. Larry Laudan, *Beyond Positivism and Relativism: Theory, Method and Evidence,* Westview, Oxford, 1997.

24. Alan Sokal, 'Transgressing the boundaries: Toward a transformative hermeneutics of quantum gravity', *Social Text,* 46/47, pp 217-252. (Reprinted in Alan Sokal and Jean Bricmont, *Intellectual Impostures,* Profile Books, London, 1998, pp 199-240).

25. Andrew Ross, *Strange Weather: Culture, Science and Technology in the Age of Limits,* Verso, London, 1991.

26. Paul Gross and Norman Levitt, *Higher Superstition: The Academic Left and its Quarrels with Science,* Johns Hopkins University Press, Baltimore, 1994.

27. Peter Coleman, 'Minority of One?' *News Weekly,* 8 June 1991, p 21.

28. Stephen Stich, cover notes for David Stove, *The Plato Cult and Other Philosophical Follies,* Basil Blackwell, Oxford, 1991.

29. Michael Levin, 'Popper and After', *Quadrant,* 27, 6, June 1983, pp 80-1.

30. David Stove, *Cricket versus Republicanism and Other Essays,* eds. James Franklin and R.J. Stove, Quakers Hill Press, Sydney, 1995.

31. David Stove, 'Cricket versus Republicanism', in Imre Salusinszky (ed.) *The Oxford Book of Australian Essays,* Oxford, Melbourne, 1997.

32. Roger Kimball, 'Who Was David Stove?' *The New Criterion,* March 1997, pp 21-8.

33. Alan Sokal and Jean Bricmont, *Intellectual Impostures,* Profile Books, London, 1998, Chapter Four.

34. David Lewis, cover notes for David Stove, *The Plato Cult and Other Philosophical Follies.*

To Chapter One

1. Feyerabend [1975], Ch. 16.
2. Kuhn [1970b] and [1970c].
3. Popper [1970] and Lakatos [1970].
4. Popper [1974b], pp. 999-1013.
5. Lakatos [1970], p. 93; Popper [1970], p. 56.
6. See, e.g., Lakatos [1970], pp. 177-80.
7. See Kuhn [1970c], pp. 236-7.
8. Feyerabend [1975], p. 167.
9. Kuhn [1970c], pp. 235-41.
10. Popper [1959], pp. 388-9.
11. Hooker and Stove [1966].
12. Popper [1959], appendix *vii; Lakatos [1968], *passim.*
13. Kuhn [1970a] , p. 121.
14. Kuhn [1970a], p. 150.
15. Feyerabend [1975], p. 27. Italics in text.
16. Feyerabend [1975], p. 39.
17. Lakatos [1970] , p. 164, footnote 1. Italics in text.
18. See, e.g., Popper [1959], p. 194.
19. Kuhn [1970a], p. 95.
20. Kuhn [1970a], p. 95.
21. Kuhn [1970b], p. 12.
22. Kuhn [1970a], p. 124.
23. Feyerabend [1975], p. 30. Italics in text.
24. Feyerabend [1975], p. 21, footnote 12.
25. Lakatos [1968], p. 347.
26. Kuhn [1970a], p. 7.
27. Lakatos [1976], p. 143.
28. Popper [1972], p. 1.
29. Popper [1972], p. 9. Cf., e.g., Popper [1963], p. 115; Popper [1972], p. 30.

30. Popper [1959], p. 104.

31. Lakatos [1968], p. 397, footnote 1.

32. Kuhn [1970a], p. 170.

33. Feyerabend [1975], p. 24.

34. Feyerabend [1975], p. 299.

35. Feyerabend [1975], pp. 301-2.

36. Feyerabend [1975], pp. 231-71.

37. e.g., Lakatos [1968], p. 383; Popper [1968], p. 293.

To Chapter Two

1. Lakatos [1968], p. 397, footnote 1.

2. Stove [1972].

3. Lakatos [1968], p. 390.

4. e.g., Bronowski [1974], pp. 615-6.

5. Popper [1957], pp. 188-9. Cf. his [1959], p. 192, starred footnote 1.

6. e.g., Popper [1959], p. 191.

7. e.g., Popper [1959], p. 196.

8. Hume [1739], p. 31.

9. Popper [1959], p. 202, footnote.

10. Popper [1959], p.191.

11. Popper [1959], p.204.

12. Popper [1959], p.204.

13. Popper [1959], p.190.

14. Popper [1959], p.192.

15. Popper [1959], p. 86. Italics in text.

16. Lakatos [1971], p. 100 and p. 125, footnote 40. Italics in text.

17. Lakatos [1970], pp. 158-9.

18. Kuhn [1970a], p. 156.

19. Kuhn [1970a], p. 159.

20. Popper [1972], p. 20. Italics in text.
21. Mill [1863], pp. 32-3.
22. Kuhn [1970c] p. 263. Italics in text.
23. Quine [1953], p. 43.
24. Lakatos [1970], p. 179, footnote 1. Italics in text.
25. Lakatos [1970], p. 109. Italics in text.
26. Kuhn [1970c], p. 238. Italics in text.
27. Lakatos [1970], p. 106.
28. Kuhn [1970c], p. 238. Italics in text.
29. Feyerabend [1975], pp. 301-3.
30. cf., e.g., Lakatos [1971], p. 126, footnote 58.
31. Kuhn [1970c], p. 264.
32. Popper [1974b], p. 1092.
33. Popper [1974b], p. 1093. Italics not in text.
34. Popper [1968], p. 297.

To Chapter Three

1. Popper [1974b], p. 1043. Italics in text. Cf., e.g., pp. 990-1, 1016, 1018-9, 1037.
2. Popper [1974b], p. 1041.
3 See, e.g., Popper [1959], pp. 29-30.
4. See, e.g., Popper [1959], p. 363 ff.
5. See, e.g., Popper [1959], p. 373.
6. Popper [1974a] , p. 69.
7. Popper [1974b], p. 1015.
8. Popper [1974b], pp. 1018-9. Italics in text.
9. e.g., Popper [1972], p. 7; Popper [1963], p. 42.
10. See, e.g., Popper [1959], p. 369; Popper [1963], pp. 42-6; Popper [1972] ,pp. 3-8.
11. The subtitle of Popper [1963].
12. See, e.g., Popper [19591, index s.v. 'Carnap'; Popper [1963],

p. 280 ff; Popper [1968], pp. 130-9, 285-303.

13. See, e.g., Popper [1963], p. 37.

14. See Popper [1959], Ch. VI.

15. Popper [1972], p. 88.

16. Popper [1972] , p. 86.

17. See, e.g., Popper [1974b], p. 1026.

18. e.g., Popper [1972], p. 89.

19. e.g., Popper [1959], pp. 369-70.

20. See Popper [1963], esp. pp. 30-6.

21. cf. Russell [1946], pp. 698-9.

22. See Popper [1972], p. 5.

23. cf., e.g., Popper [1972], p. 4.

24. Feyerabend [1975], p. 221.

25. Feyerabend [1975], p. 222. Italics not in text.

26. Kuhn [1977], p. 332.

27. See Hume [1779], p. 132.

28. Diogenes Laertius [Lives], Vol. II, p. 475; cf. Beattie [1770], p. 291.

29. Feyerabend [1975], p. 190.

To Chapter Four

1. Hume [1739], pp. 187-218.

2. Hume [1739], pp. 180-7.

3. Hume [1739].

4. Hume [1740].

5. Hume [1748].

6. Stove [1973], pp. 27-52.

7. Hume [1739], p. 139.

8. Hume [1748], p. 29.

9. Hume [1748], p. 159.

10. Hume [1748], p. 25.

11. cf. Stove [1973], pp. 35-6.
12. Hume [1748], p. 46.
13. Stove [1975], p. 17.
14. Hume [1748], p. 37.
15. Hume [1740], p. 15.
16. Hume [1739], p. 90.
17. Hume [1740], p. 15. Italics in text.
18. Hume [1748], p. 35.
19. Hume [1748], pp. 35-6.
20. Hume [1748], p. 36.
21. Hume [1739], Bk. I, Pt. III, Sect.VI, title.
22. Hume [1748], p. 36.
23. Mill [1843], p. 377, footnote.
24. cf. Stove [1976], pp. 54-8; and cf. above, Ch. III, Sect. 3.
25. Popper [1959], pp. 367; and cf. Stove [1973], p. 114.
26. Armstrong [1979], pp. 47-8.
27. Hume [1739], p. 139.
28. cf. Stove [1973], pp. 101-3.
29. cf. Stove [1965], *passim*.
30. See Williams [1947].
31. Miller [1949], p. 745.
32. MacIntyre [1969], p. 37.
33. Stove [1973], pp. 49-50.

To Chapter Five

1. Carnap [1950], p. 580.
2. Hempel [1965], Part 1.
3. Popper [1959], pp. 68-9. Italics in text.
4. Lakatos [1974], p. 247.
5. Popper [1974b], pp. 1004-5. Ellipsis in text.
6. Popper [1972], p. 198.

7. Popper [1972], p. 198.

8. See Feyerabend [1975], p. 35.

9. See Kuhn [1970c], p. 255.

10. cf. p. 30 of the same volume as contains Kuhn [1970c].

11. cf., e.g., Stove [1978], p.87; and Ch. II above, *passim*.

12. Popper [1959], p. 192.

13. Popper [1959], p. 192.

14. cf. Lorenz [1970], Vol. 1, index s.v. "vacuum activity".

15. Feyerabend [1975], p. 32.

16. Feyerabend [1975], p. 21, footnote 12.

17. Lakatos [1970], pp. 138, 140, 146.

18. Lakatos [1976], p. 29, footnote.

19. Popper [1959], p. 104.

20. See Hume [1742], Vol. 3, p. 154.

21. Barzun [1959], p. 119.

22. Hume [1748], p. 37.

23. Hume [1748], p. 34.

24. Hume [1739], p. 89.

25. cf. Descartes [1642], pp. 61-5.

26. cf. Sextus Empiricus [Outlines], p. 113, and *passim*.

27. Hume [1748]. From the author's 'advertisement' to the edition of 1777.

28. cf. Hume [1779], p. 132.

29. Quoted in Hume [1779], p. 77. Italics not in text.

30. See Popper and others in the volume which contains Popper [1970].

31. cf. Kuhn [1970a], pp. 206-7.

BIBLIOGRAPHY

Armstrong [1979]:   D. M. Armstrong, "Laws of Nature", *Proc Russellian Soc.*, Sydney University, vol. IV, 1979.

Barzun [1959]:   J. Barzun, *The House of Intellect*. Harper & Brothers, New York, 1959.

Beattie [1770]:   J. Beattie, *An Essay on the Nature and Immutability of Truth*, 1770. Bayes and Son, Edinburgh, 1823.

Bronowski [1974]:   J. Bronowski, "Humanism and the Growth of Knowledge", in *The Philosophy of Karl Popper*, ed. Schilpp, Open Court, La Salle, 1974.

Carnap [1950]:   R. Carnap, *Logical Foundations of Probability*, University of Chicago Press, 1950.

Descartes [1642]:   R. Descartes, *Meditations*, 1642, in Descartes, *Philosophical Writings*, ed. Anscombe and Geach, Nelson, London, 1966.

Diogenes Laertius [Lives]   Diogenes Laertius, *Lives of Eminent Philosophers*, Loeb Classical Library, 1970.

Feyerabend [1970]:   P. K. Feyerabend, "Consolations for the Specialist" in *Criticism and the Growth of Knowledge*, eds. Lakatos and Musgrave, Cambridge University Press, 1970.

Feyerabend [1975]:   P. K. Feyerabend, *Against Method*, NLB, London, 1975.

Hempel [1965]:   C. G. Hempel, *Aspects of Scientific Explanation*, Free Press, New York, 1965.

Hooker & Stove [1966]:   C. A. Hooker and D. C. Stove, "Relevance and the Ravens", *British Journal for the Philosophy of Science*, 1966.

Hume [1739]:   D. Hume, *A Treatise of Human Nature*, 1739, ed. Selby-Bigge, Oxford University Press, 1888.

Hume [1740]:   D. Hume, *An Abstract of ... a Treatise of Human Nature*, 1740, eds. Keynes and Sraffa, Cambridge University Press, 1938.

Hume [1742]:   D. Hume, *Essays, Moral, Political and Literary*, 1742, in David Hume, *The Philosophical Works*, ed. Green and Grose, London, 1882.

Hume [1748]:   D. Hume, *An Enquiry Concerning Human Understanding*, 1748, ed. Selby-Bigge, Oxford University Press, 1893.

Hume [1779]:   D. Hume, *Dialogues Concerning Natural Religion*, 1779, ed. Kemp Smith, Nelson, London, second edn., 1947.

Kuhn [1970a]:   T. S. Kuhn, *The Structure of Scientific Revolutions*, University of Chicago Press, 1962, second edn., enlarged, 1970.

Kuhn [1970b]:   T. S. Kuhn, "Logic of Discovery or Psychology of Research", in *Criticism and the Growth of Knowledge*, eds. Lakatos and Musgrave, Cambridge University Press, 1970.

Kuhn [1970c]:   T. S. Kuhn, "Reflections on My Critics", in *Criticism and the Growth of Knowledge,* eds. Lakatos and Musgrave, Cambridge University Press, 1970.

Kuhn [1977]:   T. S. Kuhn, *The Essential Tension*, University of Chicago Press, 1977.

Lakatos [1968]:   I. Lakatos, "Changes in the Problem of Inductive Logic", in *The Problem of Inductive Logic*, ed. Lakatos, North-Holland, Amsterdam, 1968.

Lakatos [1970]:   I. Lakatos. "Falsification and the Methodology of Scientific Research

Programmes", in *Criticism and the Growth of Knowledge*, eds. Lakatos and Musgrave, Cambridge University Press, 1970.

Lakatos [1971]: I. Lakatos, "History of Science and Its Rational Reconstructions", *Boston Studies in the Philosophy of Science*, Vol. 8, 1971.

Lakatos [1976]: I. Lakatos, *Proofs and Refutations*, Cambridge University Press, 1976.

Lorenz [1970]: K. Lorenz, *Studies in Animal and Human Behaviour*, Methuen, London, 1970.

MacIntyre [1969]: A. MacIntyre, "Hume on 'Is' and 'Ought'", in *The Is-Ought Question*, ed. Hudson, Macmillan, London, 1969.

Mill [1843]: J. S. Mill, *A System of Logic*, Longmans Green, London, 1843, eighth edn., 1941.

Mill [1863]: J. S. Mill, *Utilitarianism, Liberty, and Representative Government*, Everyman, 1948.

Miller [1949]: D. S. Miller, "Hume's Deathblow to Deductivism,", *The Journal of Philosophy*, Vol. XLVI, No. 23, 1949.

Popper [1957]: K. R. Popper, "Philosophy of Science:

a Personal Report", in *British Philosophy in the Mid-Century*, ed. Mace, Allen and Unwin, London, 1957.

Popper [1959]:   K. R. Popper, *The Logic of Scientific Discovery*, Hutchinson, London, 1959.

Popper [1963]:   K. R. Popper, *Conjectures and Refutations*, Routledge and Kegan Paul, London, 1963.

Popper [1968]:   K. R. Popper, "Theories, Experience, and Probabilistic Intuitions", in *The Problem of Inductive Logic*, ed. Lakatos, North-Holland, Amsterdam, 1968.

Popper [1970]:   K. R. Popper, "Normal Science and its Dangers", in *Criticism and the Growth of Knowledge*, eds. Lakatos and Musgrave, Cambridge University Press, 1970.

Popper [1972]:   K R. Popper, *Objective Knowledge*, Oxford University Press, 1972.

Popper [1974a]   K R. Popper, "Autobiography", in *The Philosophy of Karl Popper*, ed. Schilpp, Open Court, La Salle, 1974.

Popper[1974b]:   "Replies to My Critics", in *The Philosophy of Karl Popper*, ed. Schilpp, Open Court, La Salle, 1974.

Quine [1953]:   W.V. Quine, *From a Logical Point of View*, Harper Torchbooks, New York 1953.

Russell [1946]   B. Russell, *A History of Western Philosophy*, Allen and Unwin, London, 1946.

Sextus Empiricus [Outlines] Sextus Empiricus, *Outlines of Pyrrhonism*, Loeb Classical Library, 1967.

Stove [1965]:   D. C. Stove, "Hume, Probability, and Induction", *The Philosophical Review*, Vol. LXXIV, No. 2, 1965.

Stove [1972]:   D. C. Stove, "Misconditionalisation", *The Australasian Journal of Philosophy*, Vol. 50, No. 2, 1972.

Stove [1973]:   D. C. Stove, *Probability and Hume's Inductive Scepticism*, Oxford University Press, 1973.

Stove [1975]:   D. C. Stove, "Hume, the Causal Principle, and Kemp Smith", *Hume Studies*, Vol. I, No. 1, 1975.

Stove [1976]:   D. C. Stove, "Why Should Probability be the Guide of Life?", in *Hume: a Re-Evaluation*, eds. Livingston and King, Fordham University Press, 1976.

Stove [1978]:   D. C. Stove, "Popper on Scientific Statements", *Philosophy*, Vol. 53, 1978.

Williams [1947]:   D.C.Williams, *The Ground of Induction*,
Harvard University Press, 1947.

INDEX

(The names included in this index are those other than Popper, Kuhn, Lakatos, Feyerabend and Hume, which are mentioned in the text.)